Y0-BXQ-627

Instrumental Organic Elemental Analysis

THE ANALYSIS OF ORGANIC MATERIALS

An International Series of Monographs

edited by R. BELCHER and D. M. W. ANDERSON

1. S. VEIBEL: The Determination of Hydroxyl Groups. 1972.
2. M. R. F. ASHWORTH: The Determination of Sulphur-containing Groups. *Volume 1*, The analysis of sulphones, sulphoxides, sulphonyl halides, thiocyanates, isothiocyanates; and isocyanates. 1972. *Volume 2*, Analytical methods for thiol groups. 1976. *Volume 3*, Analytical methods for sulphides and disulphides. 1977.
3. F. KASLER: Quantitative Analysis by NMR Spectroscopy. 1973.
4. T. R. CROMPTON: Chemical Analysis of Organometallic Compounds. *Volume 1*, Elements of groups I–III. 1973. *Volume 2*, Elements of groups IVA–B. 1974. *Volume 3*, Elements of group IVB. 1974. *Volume 4*, Elements of group V. 1975. *Volume 5*, Elements of groups VIA, VIB, VIIA, VIIB, Aluminium and Zinc. 1977.
6. K. MÜLLER: Functional Group Determination of Olefinic and Acetylenic Unsaturation. *Translated from the German by* M. R. F. ASHWORTH. 1975.
7. L. C. THOMAS: Identification of Functional Groups in Organophosphorus Compounds. 1975.
8. N. K. MATHUR and C. K. NARANG: Determination of Organic Compounds with N-Bromosuccinimide and Allied Reagents. 1975.
9. E. SAWICKI and C. R. SAWICKI: Aldehydes—Photometric Analysis, *Volume 1*. 1975; *Volume 2*. 1975; *Volume 3*. 1976; *Volume 4*. 1976.
10. T. S. MA and A. S. LADAS: Organic Functional Group Analysis by Gas Chromatography. 1976.
11. J. BARTOS and M. PESEZ: Colorimetric and Fluorimetric Analysis of Steroids. 1976.
12. H. E. MALONE: The Analysis of Rocket Propellants. 1976.
13. R. BELCHER (Ed.): Instrumental Organic Elemental Analysis. 1977.

INSTRUMENTAL ORGANIC ELEMENTAL ANALYSIS

Edited by

R. BELCHER
Department of Chemistry,
The University of Birmingham, Great Britain

1977

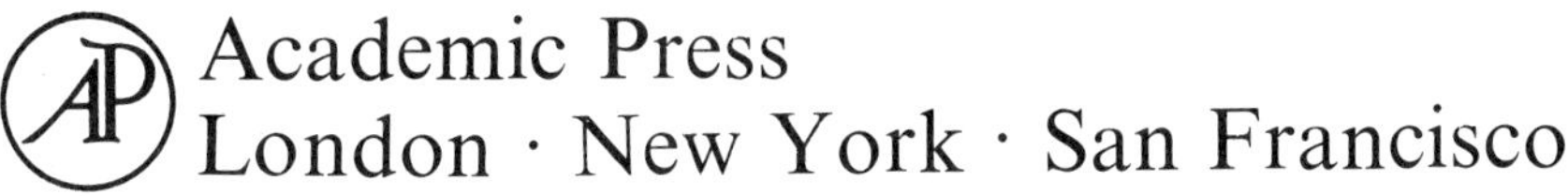

Academic Press
London · New York · San Francisco

A Subsidiary of Harcourt Brace Jovanovich, Publishers

ACADEMIC PRESS INC. (LONDON) LTD.
24/28 Oval Road
London NW1

United States Edition published by
ACADEMIC PRESS INC.
111 Fifth Avenue
New York, New York 10003

Library of Congress Catalog Card Number: 7771809
ISBN: 0-12-085950-5

PRINTED IN GREAT BRITAIN BY
Page Bros (Norwich) Ltd, Norwich

CONTRIBUTORS

P. I. BREWER, Esso Research Centre, Abingdon, Oxfordshire, England.

F. H. COTTRELL (née VAN DER VOORT) Roche Products Ltd., Welwyn Garden City, Hatfield Polytechnic, Hatfield, Hertfordshire, England.

M. R. COTTRELL, Roche Products Ltd., Welwyn Garden City, Hatfield Polytechnic, Hatfield, Hertfordshire, England.

C. J. HOWARTH, I.C.I. Ltd., Pharmaceuticalo Division, Alderley Park, Macclesfield, Cheshire, England.

F. C. A. KILLER, Esso Research Centre, Abingdon, Oxfordshire, England.

R. KUBLER, Ciba-Geigy Limited, Basle, Switzerland.

FOREWORD

Throughout the history of elemental organic analysis it has been possible to discern definite stages in its progress; during some periods, different developments have run on parallel courses.[1,2]

The first stage probably lasted for over 100 years; major changes were not forthcoming until about the beginning of the present century. Although the basic Liebig method was to continue in use until well after the second World War, the first outstanding changes were due to Dennstedt; [1,2] his methods, though not widely used, continued well into the 1930's.

In the 1920's the methods of ter Meulen were developed but, like those of Dennstedt, they were not used as widely as they deserved. Nevertheless, they have been used continuously for the determination of nitrogen, halogens, and sulphur in special cases; more recently, they have been recommended for use in conjunction with the coulometric determination. The micro methods of Pregl were somewhat later than the Dennstedt methods and overlapping with the ter Meulen methods. From the micro methods, semi-micro methods were developed, much to Pregl's disgust as indicated by the remarks in some of the earlier editions of his book. These semi-micro methods were developed by investigators such as Dubsky, Bobranski and Sucharda, Roger and MacKay and others. As it turns out, the semi-micro methods are now rarely used, but at that time, despite Pregl's tirades, there were very good grounds for their development. They did not need the specialized skill of the micro-method, an ordinary balance could be used and, very often, the standard instruments of conventional quantitative analysis, e.g. macro-burettes, were suitable. Before the great developments in instrumental analysis which changed the teaching of practical chemistry extensively, the semi-micro methods were far preferable for student teaching purposes. Especially important was the fact that the (comparatively) expensive micro-balance and micro-equipment were not essential, as universities and colleges had to work on a shoestring until well after the end of the Second World War.

In the 1945–1960 period, the "empty-tube" method was used extensively in Britain; and on the Continent of Europe active catalytic fillings, such as cobaltocobaltic oxide, and activated manganese dioxide plus silver, were favoured.

The stage with which we are now concerned is that of the automated analysers, which first began to appear in the early 1960's. This stage will obviously continue, with minor changes, into the foreseeable future.

It is, however, pertinent to mention that the idea of minimising the amount of manual labour required is not new. As early as 1905, Pregl introduced a moveable burner incorporating a clockwork motor for combustion analysis. For some years before the Second World War, partly-automated equipment was available; amongst these may be mentioned the methods of Reihlen and Weinbrenner and Bobranski and Sucharda. Automated combustion had become widespread throughout war-time Germany, as evidenced by the information in the B.I.O.S. reports;[(3)] notable contributors had been Zimmermann, Unterzaucher and Fischer. Furnaces operated by synchronized motors became commonplace in Britain after the War; (they were, however, only of use in slow combustion processes).

The main feature of the new automated methods, at least initially, was the elimination of the weighing procedure. Very fast combustion methods of various kinds had been developed long before, but, despite the introduction of various devices to accelerate weighing, this operation remained the main hindrance to reducing the time of operation further. The high sensitivity of the new instrumental forms of final measurement enabled a considerable reduction to be made in sample size; the more rapid combustion now possible is due to the small amount of sample taken and not to any great improvement in combustion efficiency.

The two systems chosen for inclusion in this Monograph are those of Perkin–Elmer and Carlo Erba, for these are the two most widely used. The former, by suitable modification, can also be used for the determination of oxygen and sulphur and the latter also for the determination of oxygen.

The determination of carbon in waters has become of great importance; several analyzers based on different principles have been developed commercially. It was considered that a general account of the available equipment would be of value in this Monograph. One feature which should interest older analysts is the system in which the combustion product, carbon dioxide, is converted to methane for measurement; in former times the aim was always to convert methane to carbon dioxide.

Undoubtedly, the simultaneous determination of nitrogen is one of the major advantages of the automatic methods. Nevertheless, the situation where only the nitrogen content is required often arises and nitrogen may be only a minor constituent, in which case the automatic method may not provide the accuracy required. The best of the newer systems, that of Merz, can be applied to the analysis of virtually any type of organic material and to any range of nitrogen. A short account of this method has therefore been included.

Finally, the great developments in coulometric analysis have led to the wide application of this technique in elemental analysis; all the common organic elements (other than carbon and hydrogen) can be determined. Again, the great sensitivity of detection allows very small samples to be taken and this, in turn, decreases the time required for decomposition.

These methods of elemental organic analysis based on instrumentation have been well-established for more than a decade. As far as it is known, no monograph has dealt exclusively with these methods. It was felt that such a monograph would be useful to the increasing number of laboratories using such apparatus. I was fortunate in obtaining the services of analytical chemists as contributing authors who have had first-hand experience with such systems for a large number of years. It is hoped that this monograph will be of benefit to all analysts who need to apply these methods.

References

1. R. Belcher, *Proc. Analyt. Div. Chem. Soc.*, 1976, **13**, 153.
2. *idem, ibid.*, 1975, **12**, 77.
3. B.I.O.S. Report 715, 1946.
   B.I.O.S. Report 1606, 1948.

Birmingham
July 1977

R. Belcher

CONTENTS

1. THE PERKIN–ELMER MODEL 240 ELEMENTAL ANALYZER: CARBON, HYDROGEN AND NITROGEN

M. R. COTTRELL and F. H. COTTRELL (neé: van der Voort)
Roche Products Ltd, Welwyn Garden City,
and
Hatfield Polytechnic, Hatfield, Hertfordshire

I. INTRODUCTION

During the early 1960s, details were published[1] of a new approach to organic elemental microanalysis. Professor W. Simon and co-workers of the Swiss Federal Institute of Technology in Zürich showed that it was possible to automate some existing procedures of microanalysis, in particular carbon and hydrogen, and as a result many of the drawbacks of the classical methods could be eliminated.

On the basis of Professor Simon's original work, the Perkin–Elmer Corporation of Norwalk, Connecticut, U.S.A., developed in 1963 an instrument for the automatic determination of carbon, hydrogen and nitrogen, in which the classical techniques of Pregl for carbon and hydrogen, and Dumas for nitrogen, were combined. Many of the manual sequences of the classical methods, such as the use of bunsen burners, weighing of absorption tubes, etc., were rendered out of date and replaced by automatic combustion and modern electronic detection of the gaseous products. The combustion system of the instrument differs partly from the classical methods, for its sample combustion takes place under static conditions in a constant amount of oxygen, consequently the blank values are not affected on prolonged combustion of refractory compounds. Further application of modern techniques resulted in an automatically controlled and versatile instrument, which provided excellent reproducibility of the conditions, while variability of these conditions was incorporated to suit the type of compound to be analyzed.

The Model 240 elemental analyzer (Fig. 1) was introduced commercially in 1965[2] and during the following years of continued research by the Company, only small changes, such as updating of electronic components, were made to the instrument itself. One of the developments was a conversion of the instrument to allow direct determination of oxygen[3] and later also sulphur;[4] these conversions will be discussed in a separate chapter.

Attention was given to further automating the various stages of operation from weighing of the sample to the final reporting of results. Most advances

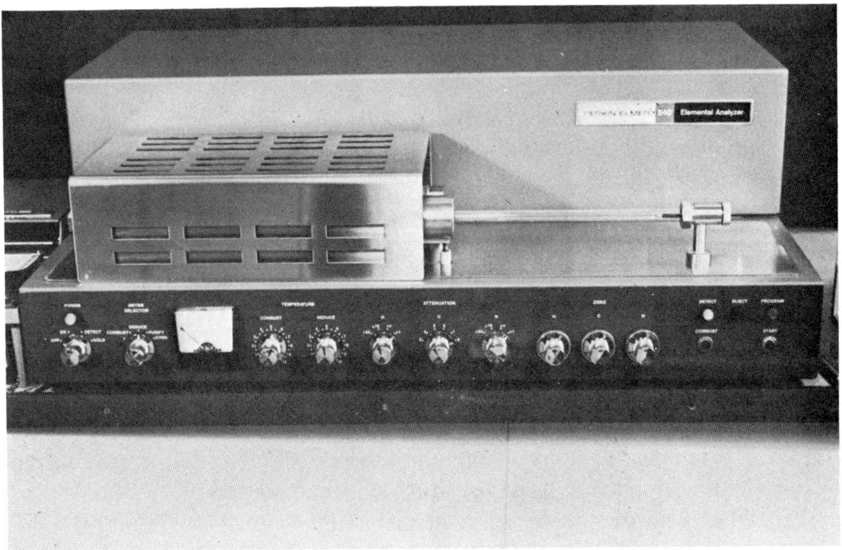

FIG. 1. Model 240 Elemental Analyzer.

in automation were made in those areas where reproducible operation from one analysis to another could be applied without the need for operator involvement. Recently[5] an automated sample introduction system, capable of handling up to 60 samples, was introduced and its combination with a programmable calculator rendered the system almost fully automatic.

The stage of sample preparation, however, is dependent on the type of compound to be analyzed. Compounds, which are for example a liquid, air-sensitive, or boron derivatives, require special treatment in the preparation. Automation of the sample preparation stage may be feasible when relatively simple and stable compounds, which require little more than accurate weighing, are to be analyzed continuously.

As in most microanalytical laboratories a variety of compounds are handled, most stages, except the sample preparation, can at present be made automatic by means of accessories to the Model 240 elemental analyzer.

Although the instrument is at present predominantly used for routine analysis in organic chemistry, the wide variety of compounds analyzed show that its field of application is very wide.

Weighing is an important part of an accurate analysis and as sophisticated automatic instrumentation cannot be justified without an accurate balance, some mention will be made of suitable balances.

The Model 240 microanalyzer takes generally 13 minutes for the completion of one analysis, consequently a balance capable of weighing milligram levels of sample to an accuracy of 0·1% within a few minutes is essential for the continuous operation of the instrument. Several modern electrobalances as well as precise knife-edge balances satisfy the above requirements.

Certain changes in environmental conditions, such as humidity, temperature, draughts, and vibration, may influence the performance of the balance as well as the Model 240 analyzer; their effects will be mentioned later (p 48).

Details on planning of a microlaboratory have been published,[(6)] but some suggestion will be made in connection with the siting of the Model 240 and its accessories.

In order to avoid spillage on transfer of the sample, the balance is best positioned to the right of the instrument, preferably on an anti-vibration bench, and a small working area between the balance and the instrument is useful for preparation of the sample before its introduction into the Model 240. The data handling device is best situated to the left of the instrument within easy viewing distance of the operator in case attenuation is required. A calculator should be at hand for immediate calculation of the results.

These chapters on the Model 240 are mainly aimed at providing basic information to students and present or future users of the instrument. In order to promote the understanding of basic functions of the instrument, some detailed information will be included, but it is not intended to cover in detail aspects which are readily available in the Perkin–Elmer manual.[(7)]

II. PRINCIPLES OF AUTOMATION AND CHEMICAL REACTIONS

A divison has been made here between the basic stages and the automatic sequences of the Model 240, in order to assist in the understanding of its operation.

The basic stages cover briefly all stages of the analysis, while in the automated sequences the stages are explained in more detail in connection with the automated operation of the instrument.

The chemical reactions of combustion, reduction, and absorption are discussed separately.

A. Basic stages of operation

The operating functions in connection with the Model 240 can be divided into five main stages, as shown in Fig. 2.

1. *Sample preparation*

This stage is dependent on the particular methods of sample preparation

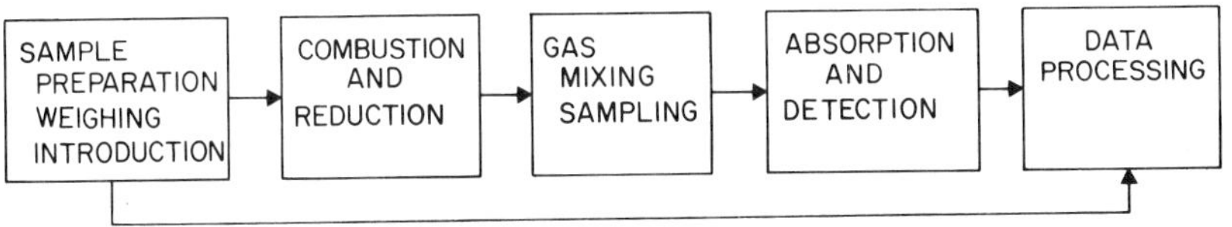

FIG. 2. Operating stages of the Model 240.

employed, such as drying, special weighing techniques (encapsulation, etc.), and addition of oxidants, but basically the accurately weighed sample is transferred to a sample carrier for convenient introduction into the instrument.

2. *Combustion and reduction*

The instrument is prepared for sample combustion by introducing pure oxygen into the combustion tube. The combustion furnace is held at a temperature of approximately 950°C. After oxidation of the sample, interfering compounds are removed and the products H_2O, CO_2, and oxides of nitrogen are transferred to the reduction stage by means of helium. The oxides of nitrogen are reduced to elemental nitrogen and residual oxygen is removed. The final combustion products of H_2O, CO_2 and N_2 are transferred by means of helium to the mixing stage.

3. *Gas mixing and sampling*

The four gaseous components are homogenized in a constant volume, temperature and pressure and then passed through a sampling volume before part of this mixture is used for detection.

4. *Absorption and detection*

The gaseous mixture from the sampling volume is transferred to the detectors where chemical absorbers remove sequentially H_2O and CO_2. A detection system measures conductivity changes before and after absorption and the differential signal for each gas is recorded.

5. *Data processing*

The signals from the detectors are normally displayed on an analog strip chart recorder, but other more sophisticated data-handling devices such as printers and computers can be used. The recorded data are then processed to give the percentages of carbon, hydrogen and nitrogen. Information from a balance can also be fed in at this stage, when computing devices are employed.

This completes the analysis and the instrument automatically returns to a standby condition awaiting the introduction of a further sample.

B. Automated sequences of operation

The automated operation of the Model 240 is controlled by a precise programmer unit which completes a total of 64 switching functions at predetermined times during a 13-minute analysis cycle. High purity helium and oxygen are used for purging and oxidation respectively, while the pressure is controlled via stainless steel diaphragm regulators which have a very low rate of diffusion. Highly reproducible inlet pressure control is essential, particularly in connection with the helium supply, if the desired analytical precision is to be achieved. In order to avoid slight changes in pressure and volume within certain critical parts, such as mixing and sampling volumes, pressure switches, and detectors, these parts are enclosed in an oven which is maintained at 75° ± 0·05°C.

Gas flow rates and directions are controlled by capillary restrictor tubes and solenoid valves. The schematic diagram of Fig. 3 will be used to describe the operating sequence together with the gas flow directions at each stage of the operation.

1. *Standby condition*

The Model 240 is maintained in a standby condition awaiting the start of

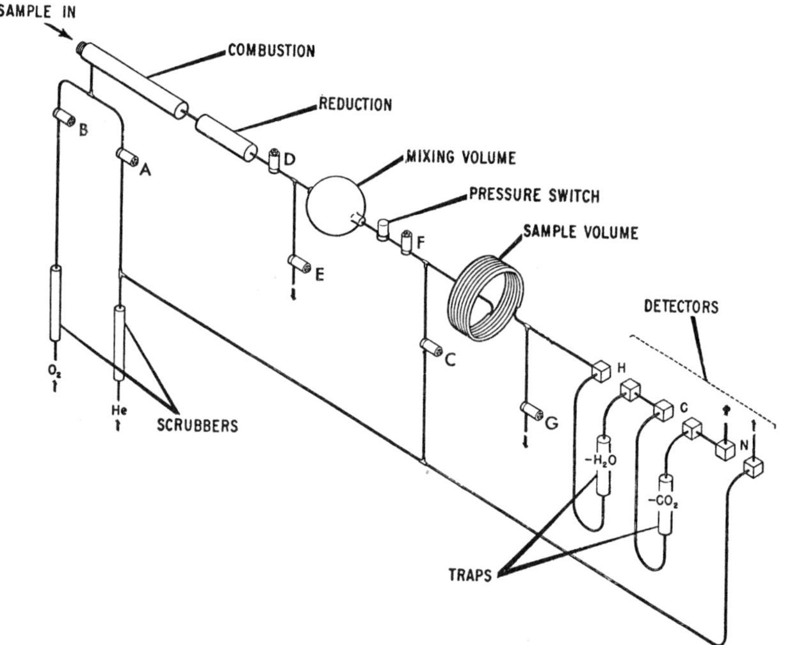

FIG. 3. Schematic diagram of the instrument.

an analysis. Helium from which H_2O and CO_2 are removed by scrubbers continually purges the vessels of the instrument at a flow rate of approximately 30 ml/min. The route which is followed by helium through the instrument in standby condition is shown in Fig. 4. The combustion and reduction furnaces are maintained at their normal operating temperatures of 950°C and 670°C respectively.

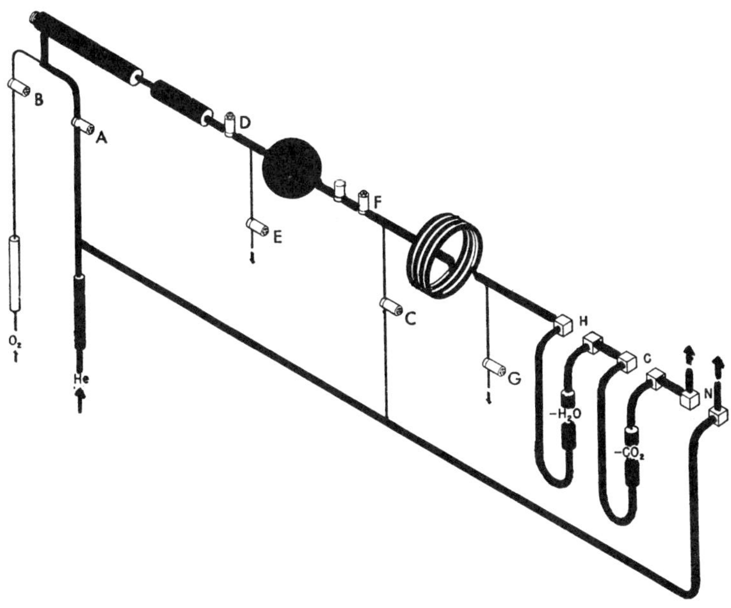

FIG. 4. Flow of helium in stand-by condition.

2. *Sample introduction and purging sequences of gases*

At the start of an analysis a weighed sample is introduced into the cold section of the combustion tube by means of a quartz sample carrier and the program is initiated by a push button on the front panel of the instrument. Initially all solenoid valves are opened (except the oxygen valve, valve B), which allows a purging sequence by helium through all the vessels of the instrument at various flow rates, as shown in Fig. 5. Air introduced with the sample, the carrier, and any remaining combustion gases from the previous analysis are removed from the system before the combustion of the new sample takes place.

After having purged with helium for a period of $1\frac{3}{4}$ minutes, valve A closes which stops the helium flow to the combustion and reduction tubes, and valve B opens allowing pure oxygen to enter the combustion tube; this is

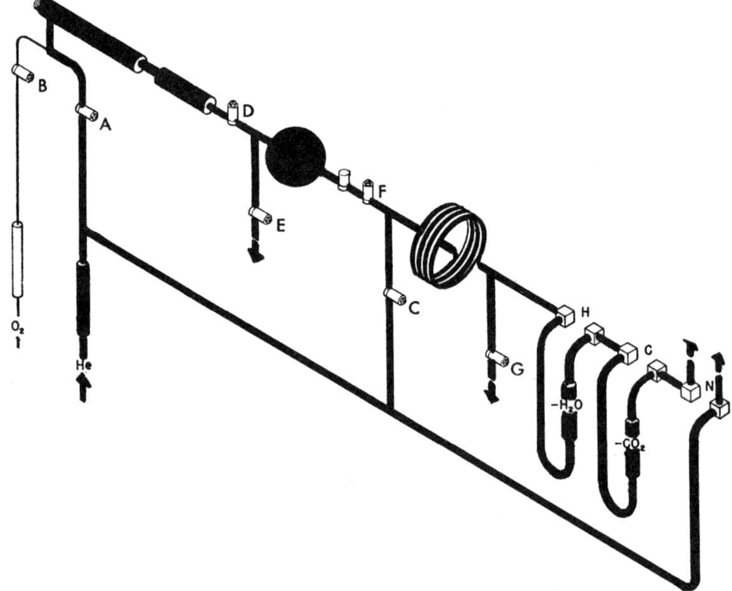

FIG. 5. Flow of helium in the purge condition.

shown in Fig. 6. Helium continues to purge through the mixing volume, sample volume and detectors, in order to remove all residual gases from the system. Valve B remains open for 30 seconds, during which time about 25 ml of oxygen enters the combustion tube. This amount of oxygen can be varied by adjustment of the inlet gas pressure, and the flow can be checked at the sample entry port. When valve B closes, the electronic programmer stops, which is indicated by a light on the front panel informing the operator that the sample can be introduced into the combustion furnace.

3. *Combustion and reduction sequences*

The sample carrier is transferred to the combustion furnace by means of a simple hand-held magnet, which positions the sample in the hottest part of the furnace. This operation can be automated by means of a recent accessory to the Model 240 (p. 23).

As the combustion takes place under static conditions, valves D and E remain open for the first 30 seconds of the combustion time, which allows relief of pressure to atmosphere, in order to avoid damage due to sudden expansion of gases when highly reactive samples are analyzed. The remaining $1\frac{3}{4}$ minutes of combustion is sufficient to oxidize the majority of compounds, but the combustion time can be extended by holding the program cycle for any length of time.

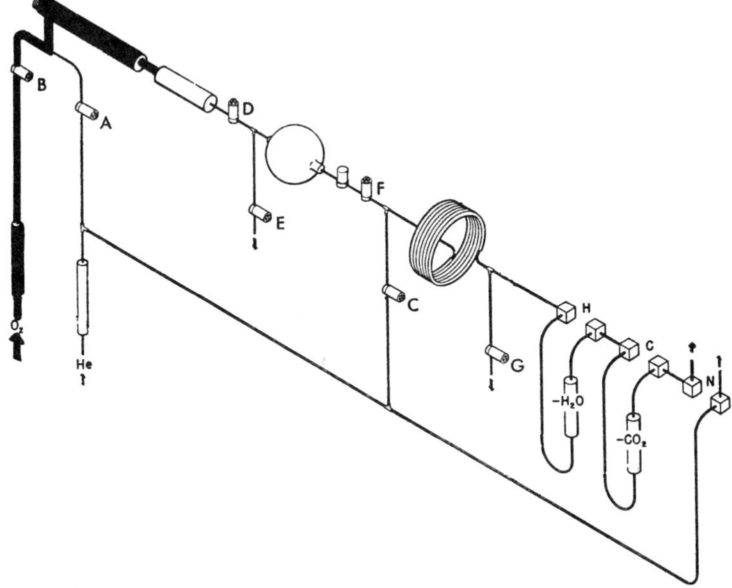

FIG. 6. Introduction of oxygen into the combustion tube.

The last seven inches of the combustion tube contain a series of chemicals, some of which assist in the oxidation and some absorb interfering combustion products (p. 26). At the end of the normal combustion period a flash heater, which is positioned just before the combustion furnace, is switched on and its temperature of approximately 1000°C quickly vaporizes any condensates. Then those oxidation products, which are not absorbed, are carried to the reduction tube by means of helium. The hot copper in the reduction tube removes any remaining oxygen and reduces the oxides of nitrogen to elemental nitrogen.

4. *Homogenizing of the combustion products*

During the final part of the oxidation–reduction stage valves E and F close, so that the remaining H_2O, CO_2, N_2 and helium flow into the mixing volume (Fig. 7). When the helium pressure reaches 1500 mm Hg, a sensitive pressure switch operates, which closes valve D and encloses the three combustion gases and helium in the mixing volume (Fig. 8). This condition is maintained for a period of $1\frac{1}{2}$ minutes in which time a homogeneous mixture of the four gases is formed.

At the same time signals from the three detector bridges are recorded for

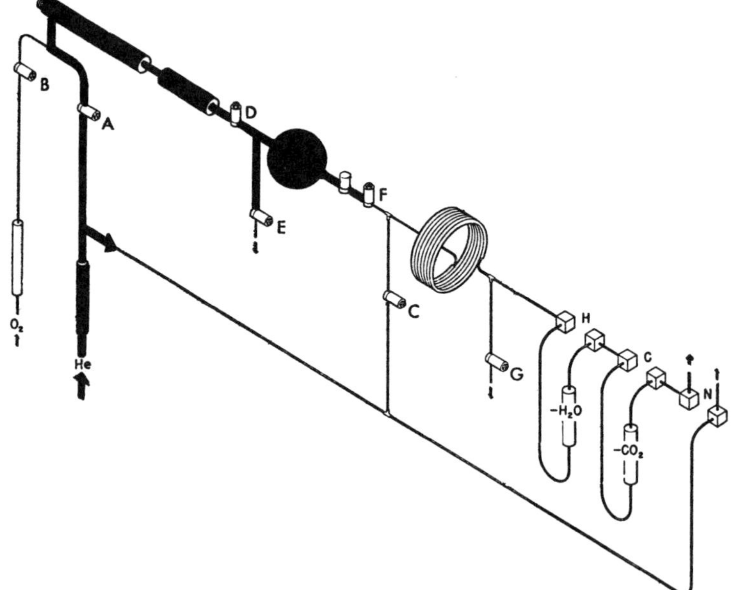

Fig. 7. Transfer of gaseous products to the mixing volume.

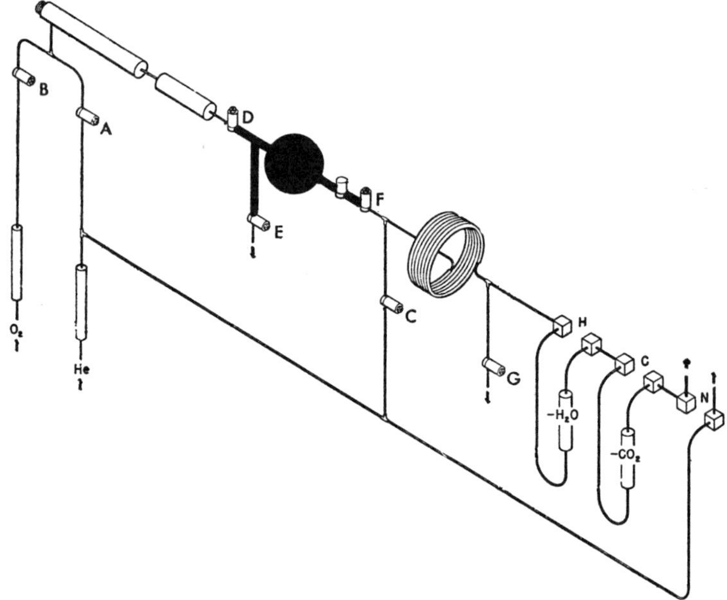

Fig. 8. Mixing of the gaseous products

a period of 20 seconds each (Fig. 9). These pure helium signals represent a reference or zero level and are used in the final calculations.

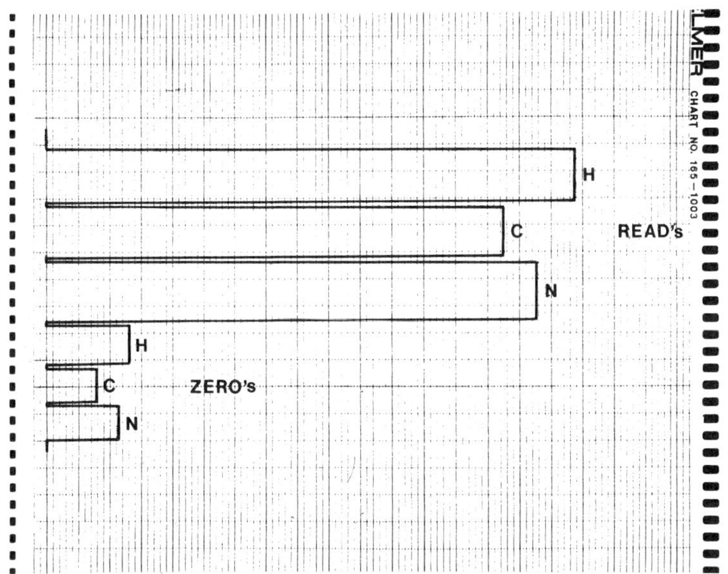

FIG. 9. Model 240 recorder readout.

5. *Sampling of homogenized combustion products*

Following the mixing stage, valve C closes and valves F and G open allowing the homogeneous gas mixture to expand through the sample volume and excess gas to flow to atmosphere. In this way the pressure is reduced to a constant level (near atmospheric pressure) and part of the gas mixture is retained in the sample volume (Fig. 10).

6. *Detection and data processing*

By closing valves G and F and opening valve C the gaseous mixture flows from the sample volume to the detectors (Fig. 11), where the products H_2O and CO_2 are absorbed. Detector filaments are situated before and after each absorption tube through which the gas mixture flows at a constant rate; the difference between the output of each set of detectors before and after absorption of each product is recorded as the read signal (Fig. 9). After removal of H_2O and CO_2 only N_2 and helium remain which are related to pure helium to give the nitrogen read signal (Fig. 9). Each read signal is recorded for a period of 30 seconds.

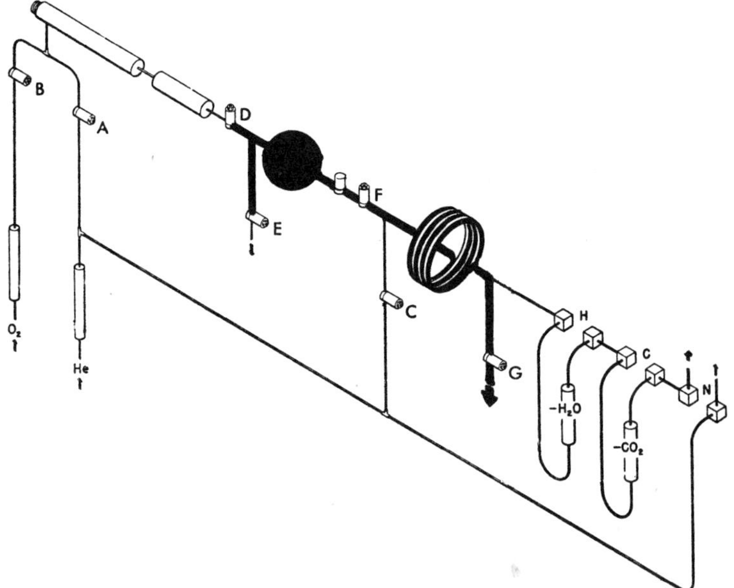

FIG. 10. Sampling of the gas mixture.

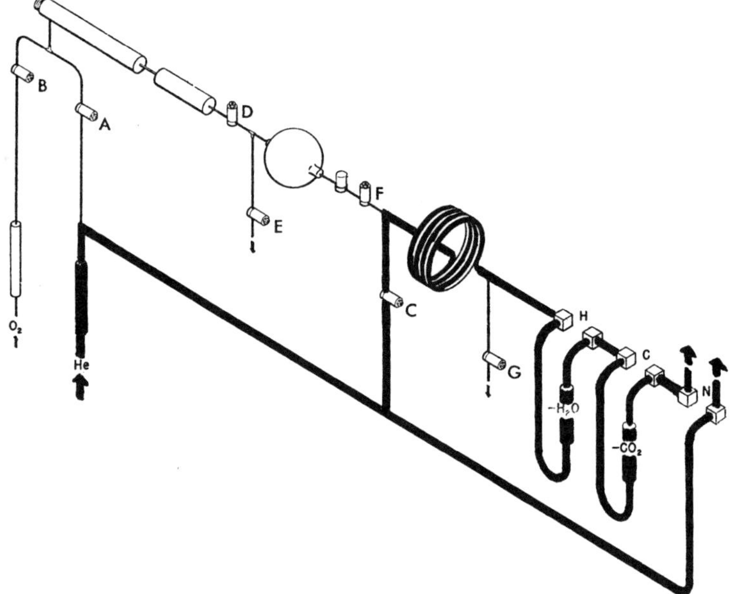

FIG. 11. Detection of the gaseous products.

On completion of the detection stage, valve C closes and valves A, D, and F open, which returns the instrument to the standby condition.

The recorded values of Fig. 9 are used to calculate the constants of the instrument in microvolts signal per microgram of element based upon a known standard compound, or alternatively these derived constants are used to calculate the percentages of carbon, hydrogen and nitrogen in the samples analyzed.

C. Chemistry of combustion, reduction and absorption

The established chemical technology of the classical methods in the field of microanalysis provided the foundation for the basic design concept of the Model 240 elemental analyzer. Accordingly, the chemistry involved in the analysis of a compound in the Model 240 is similar to that of the Pregl-Dumas methods, which have been discussed in the literature.[6,8,9] Some basic knowledge of the chemistry involved is briefly discussed here.

1. *Combustion*

The sample is decomposed in the presence of oxygen at a temperature of 950°C and converted mainly into the oxides of the various elements present in the sample. The reactions of the elements to be analyzed are as follows:

$$C \xrightarrow[O_2]{950^\circ} CO_2$$

$$H \xrightarrow[O_2]{950^\circ} H_2O$$

$$N \xrightarrow[O_2]{950^\circ} N \text{ oxides}$$

The interfering products are subsequently absorbed by certain chemicals used in the packing of the combustion tube and the remaining products H_2O, CO_2, oxides of nitrogen and residual oxygen are transferred by means of helium to the reduction stage.

2. *Reduction*

The reduction tube is packed with copper and it is maintained at a temperature of 670°C. When the products H_2O, CO_2, the oxides of nitrogen and the residual oxygen are passed through this tube, the oxides of nitrogen and the oxygen are reduced as follows:

$$O_2 \xrightarrow[Cu]{670^\circ} CuO$$

$$N \text{ oxides} \xrightarrow[Cu]{670^\circ} N_2 + CuO$$

The remaining gaseous products H_2O and CO_2 are not reduced and these products together with the nitrogen formed during the reaction are subsequently transferred to the absorption stage by means of helium.

3. *Absorption*

In the final stage of the analysis, the gaseous products H_2O and CO_2 are sequentially removed by absorbers in two separate tubes.

One tube contains a water absorber, usually magnesium perchlorate, and the other a carbon dioxide absorber, usually sodium or lithium hydroxide. The carbon dioxide absorption tube also contains some magnesium perchlorate, which removes the water formed during the absorption of the carbon dioxide.

The reactions are as follows:

$$6\,H_2O + Mg(ClO_4)_2 \longrightarrow Mg(ClO_4)_2 \cdot 6\,H_2O$$

$$CO_2 + 2\,NaOH \longrightarrow Na_2CO_3 + H_2O$$

The remaining nitrogen and helium are finally purged to atmosphere.

III. APPARATUS—DESCRIPTION OF INSTRUMENTAL COMPONENTS AND ACCESSORIES

The main components of the Model 240 elemental analyzer can be divided into four groups: front panel, combustion and reduction, gas mixing–sampling and detection, and electronics and programmer. Figure 12 shows the layout of these four groups of components in the instrument.

A. Front panel controls

The front panel controls shown in Fig. 13 are used for selection, adjustment and indication of the various operational stages of the Model 240. The general functions of the controls can be divided into the following sections:

1. Power switch and light
2. Temperature read-out and select switch
3. Temperature controls for the furnaces
4. Attenuation switches for H, C, and N
5. Zero controls for H, C and N
6. Program sequence, indication lights and push buttons.

B. Combustion and reduction

The combustion and reduction tube layout in the Model 240 allows convenient maintenance; it is diagrammatically shown in Fig. 14.

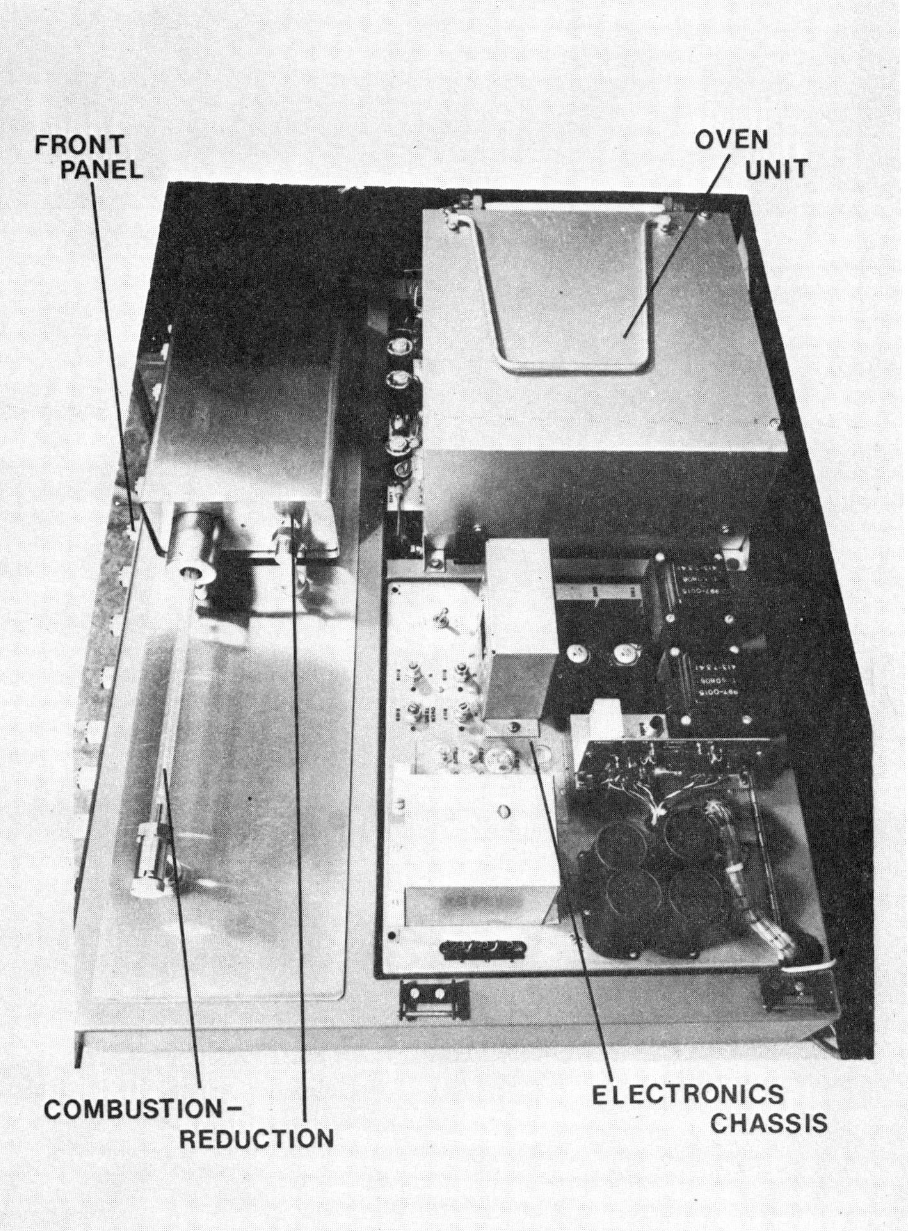

FIG. 12. Lay-out of the components in the Model 240.

FIG. 13. Front panel controls.

The sample entrance port is designed in such a way as to facilitate convenient introduction and removal of the quartz ladle; it holds the end of the combustion tube and it further provides connections to both helium and oxygen gas inlet tubes through valves A and B, which are positioned directly underneath the entrance port.

The independently heated furnaces offer temperature control from 100°C up to about 1000°C for combustion and from 100°C to 850°C for reduction, although the usual operational temperatures are respectively 950°C and 670°C.

The standard combustion tube is manufactured from high quality transparent quartz and has a total length of 25 inches, 12 inches of which are inside the furnace, while the combustion tube packings occupy $7\frac{1}{2}$ inches at the end of the heated part of the tube. A temperature gradient is applied over the furnace length to give a constant high temperature (950°C) for oxidation from the right side of the furnace to well into the packings and a gradually lower temperature towards the left to suit the silver reagents which absorb halogens and sulphur.

To the right of the combustion furnace a flash heater is located, which raises the temperature of the tube to approx. 1000°C in 30 seconds, to

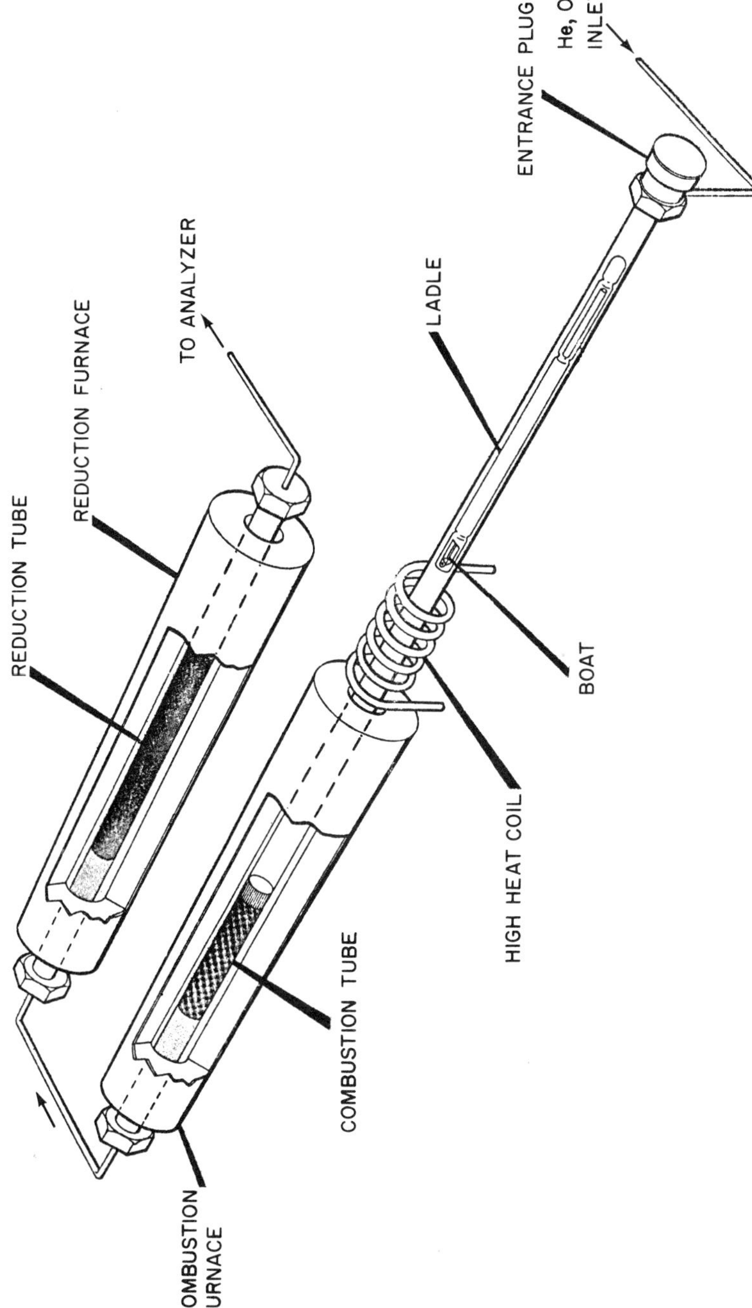

FIG. 14. Combustion and reduction tube lay-out.

facilitate rapid oxidation of condensates which may have formed during the combustion under static condition. This heater is energized just before helium gas carries the oxidation products through the packings of the combustion tube.

The reduction tube is $14\frac{1}{4}$ inches long and is manufactured from transparent quartz, normally of a lower quality than the combustion tube; it contains 12 inches of copper granules for reduction purposes together with small plugs of silver at each end. A furnace temperature of about 670°C is used for the reduction and the total tube length is held at this temperature although some decrease in temperature is inevitable towards each end.

Recently Perkin–Elmer has introduced a larger capacity reduction tube and associated furnace; the larger amount of filling in this tube is expected to last four times longer than the filling in the conventional tube (p. 29). The furnace associated with the larger reduction tube allows inspection of the filling through a window.

The ends of both combustion and reduction tubes are made airtight by means of rubber O-rings, which can easily be removed and replaced. A stainless steel fitting is used to connect these tubes together at the left end of the furnace unit.

C. Gas mixing, sampling and detection

1. *Oven unit*

The mixing and sampling volumes, pressure switch and detector block, are located in an oven unit, which is maintained at a constant temperature of 75°C (±0·05°C), since their pressure and volumes have to be controlled in order to provide reproducible results. The absorption tubes are located outside the oven for convenient maintenance (p. 52). The oven and its contents are shown in Fig. 15.

2. *Mixing volume*

A heated restrictor tube connects valves D and E, which are situated beneath the outlet of the reduction tube, to a spherical 300 ml glass mixing volume. The inlet and outlet ports of the mixing volume are designed to give good flow characteristics both for filling and purging.

3. *Pressure switch*

In order to control the pressure within the mixing volume, a pressure switch activates valve D when 1500 mm mercury pressure is reached. The pressure switch is of a flow through type with nickel bellows acting as a pressure sensor, which, when the desired pressure is reached, operate the plunger of a microswitch thus completing a circuit with valve D. All com-

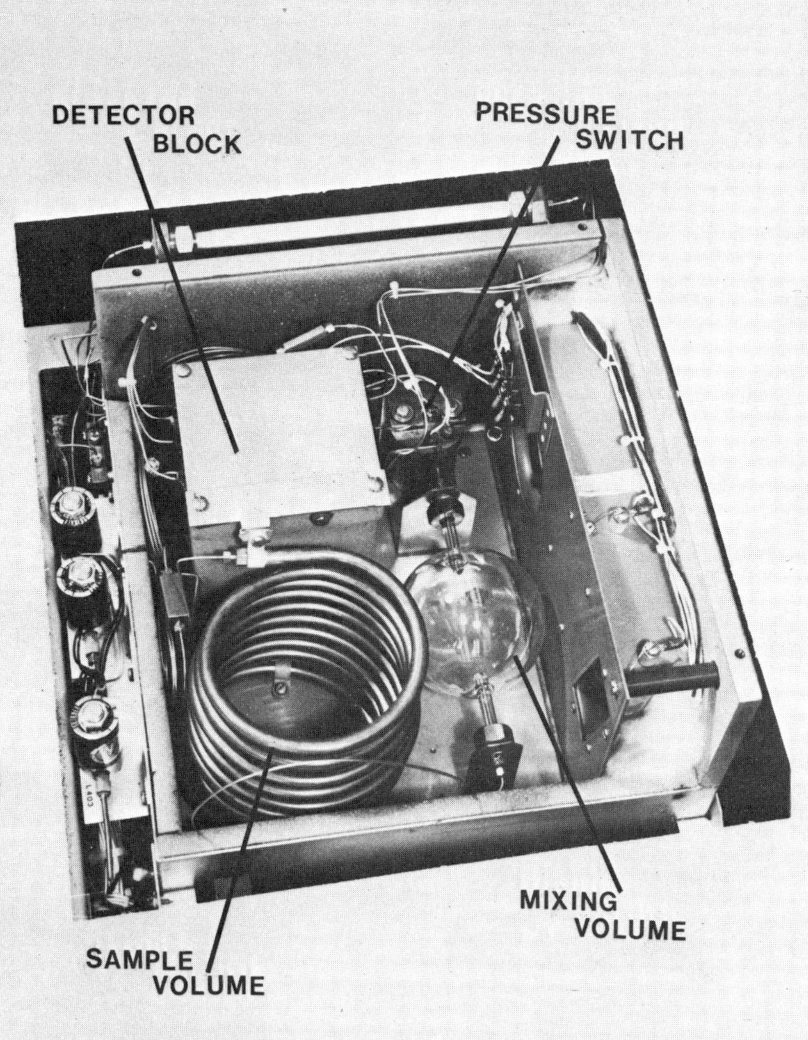

FIG. 15. Contents of oven unit.

ponents of the pressure switch are enclosed in a brass encasement, which is sealed during manufacture.

4. *Sample volume*

A capillary tube connects the outlet of the pressure switch to the inlet of the sample volume through valve F. The sample volume is a coiled copper tube of about 12 feet in length and a volume of 175 ml and it has a delaying action on the gas mixture flowing from the mixing volume. Apart from allowing the release of excess pressure to the atmosphere through valve G, it ensures that the gas mixture flows through the detectors at a constant rate and it provides a sufficient volume of homogeneous gas mixture for a 30 seconds recording of each detector bridge output.

5. *Detector block*

The outlet of the sample volume is connected to the detector block by a capillary tube. The detector block is made from aluminium and contains the three thermal conductivity cells in a stainless steel mounting. The three cells are connected in series and each cell consists of four platinum hot wire filaments, which are connected differentially in a bridge circuit as shown in Fig. 16.

The chemical absorbers are situated between each half of the water and

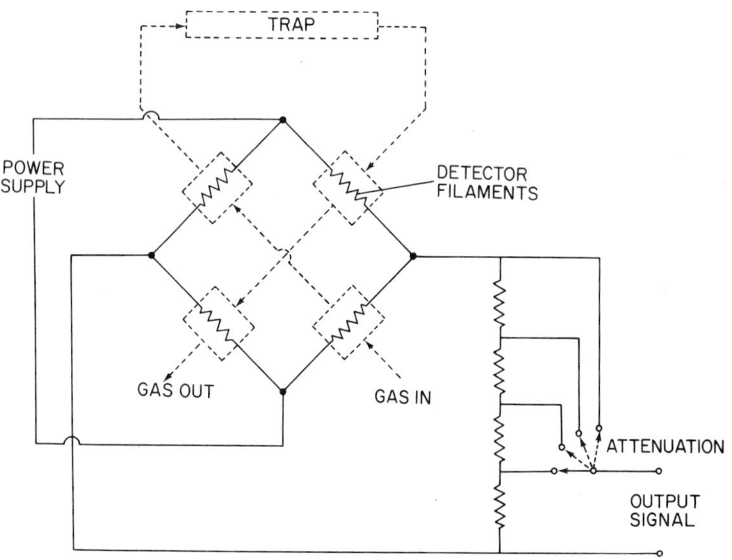

FIG. 16. Detector bridge configuration.

carbon dioxide detector bridges, in order to produce the desired differential signal. The differential signal for nitrogen is derived from nitrogen and helium passing over one half of the nitrogen detector bridge and only helium passing over the other half.

The output from each bridge is automatically recorded for a period of 30 seconds each by the programmer and they produce a series of constant signals, which are shown in Fig. 9 (p. 11). When the output from the nitrogen and hydrogen detectors is greater than the 1 mV input of the recorder, this signal can be attenuated ($\times 4$, $\times 16$, $\times 64$) by switching in fixed resistors before it is accommodated on the recorder. When analyzing organic samples, the signal level from the carbon bridge is usually higher than that of either the nitrogen or hydrogen bridges, therefore the signal is divided by ten through a fixed attenuation; further suppression of the signal can be provided manually by off-set voltages (1–5 mV). Consequently the precision of the final read-out remains high for all three elements.

6. *Chemical absorbers*

The absorbers are contained in glass tubes, which are located outside the oven unit for easy maintenance and in close proximity to the detectors. Rubber O-rings are used for airtight sealing and they are identical to those used in the combustion and reduction tubes.

The packings of the absorption tubes will be discussed later (p. 29).

D. Electronics, programmer, and solenoid valves

1. *Electronics*

Solid state electronic components are employed in the instrument; they are highly reliable and therefore require minimal attention during normal operation. The main components control the power to the detectors, solenoid valves and furnaces, while an inbuilt safety circuit protects the detector filaments from overheating. A stabilized power supply is used for the detectors (18 V d.c.) and the solenoid valves (24 V d.c.).

2. *Safety circuit*

The changes in the conductivity of the gas flow are indicated by the platinum hot wires of the detector bridges, which may overheat when in contact with oxygen. Therefore a safety circuit is incorporated in the nitrogen bridge, which automatically switches off the power supply to all detectors and is indicated by a red light on the electronic chassis.

The power supply to the detectors can only be restored by manually operating a button after the oxygen has been removed by helium.

3. *Programmer*

The timing sequence of the instrument is controlled by a 20-position programmer, which reproduces a program to ±0·25%. A motor-driven camshaft activates microswitches, which control the various operations throughout the 13-minute cycle of analysis.

4. *Solenoid valves*

As the seven solenoid valves of the Model 240 open and close in total 24 times during the 13-minute cycle of analysis, reliability of operation is of importance, which has been achieved by their relatively simple design as shown in Fig. 17. The only moving parts in the solenoid valves are the plunger and the spring, which occasionally need replacing as well as the O-rings sealing the valves.

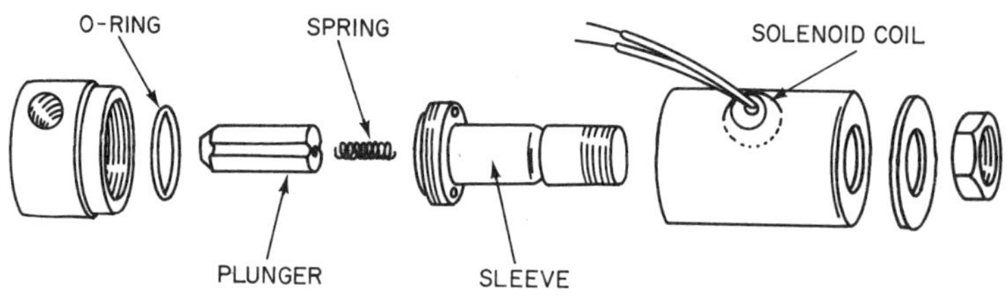

FIG. 17. Components of the solenoid valve.

E. Accessories

1. *Pre-purifier*

Purification of the helium and oxygen gases is not necessary if high-purity gases (research grade) are employed, which have a low level of carbon dioxide, water and nitrogen. When analyzing very low levels of elements or if pure gases are not available, purification of the gases can be achieved by a pre-purifier, such as the Perkin–Elmer accessory shown in Fig. 18. The helium and oxygen gases are passed through two separate reactor tubes, which are situated in a double furnace unit and are maintained at 625°C. The chemicals used as packings in these tubes determine which impurities will be removed from the gases, and they will be discussed later (p. 30). The tubes are sealed by means of rubber O-rings and pressure control is obtained by separate regulators and gauges. All electrical connections are made to terminals on the Model 240.

2. *Automatic sample introduction and data handling system*

The automation of the sample introduction has received attention from users as well as manufacturers and some different designs have evolved.

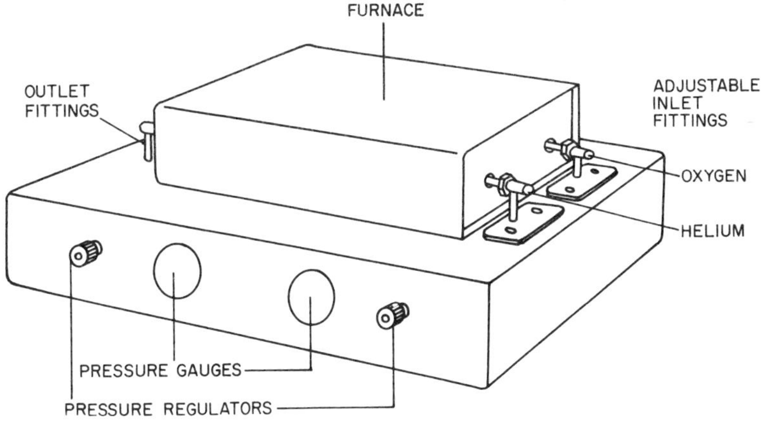

FIG. 18. Perkin–Elmer gas pre-purifier.

(a) *Single sample introduction.* Some users[10] have built single sample introduction systems, which are based on the automatic transfer of the sample ladle from outside to inside the combustion furnace and vice versa. Since these devices introduce one sample at a time, they provide only a marginal improvement over the manual method.

(b) *Multi-sample introduction.* Perkin–Elmer has recently[5] introduced a sample introduction accessory, which handles up to 60 samples and is combined with a data handling system to provide fully automated operation. Either the data handling system or both accessories can be fitted directly to any Model 240. Figure 19 shows the basic design of this multi-sample

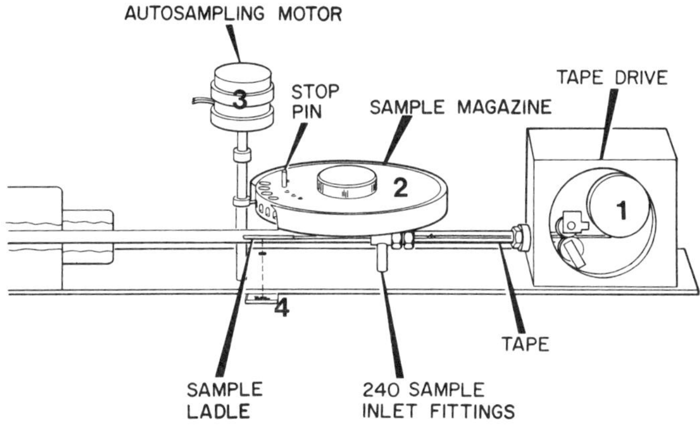

FIG. 19. Multi-sample introduction system by Perkin–Elmer.

introduction system. The ladle is coupled to a motor-driven tape (1), which gives forward and reverse motion. Up to sixty samples in their separate containers are placed in a circular magazine (2) and an autosampling motor (3) initiates the transfer of the containers from the magazine to the ladle. The retrieved containers are ejected into a tray (4) underneath the combustion tube.

The various operations of the autosampling system are controlled by an encoder attached to the normal programmer unit of the Model 240. This encoder can be modified to suit oxygen or sulphur analysis.

A safety circuit is incorporated, which switches the autosampler off in events such as a power failure or when the light of the detector safety circuit is on.

(c) *Automatic data handling*. The data handling system, which is combined with the multi-sample introduction system described above, has three main functions:

(1) control the analysis of up to 60 samples by means of the auto sampler,
(2) processing of all data and calculations,
(3) printing an analytical report of the results.

A typical printed report of a CHN analysis is shown in Fig. 20, and it can

TYPICAL ANALYTICAL REPORT

Printout

Cystine $C_6H_{12}O_4N_2S_2$

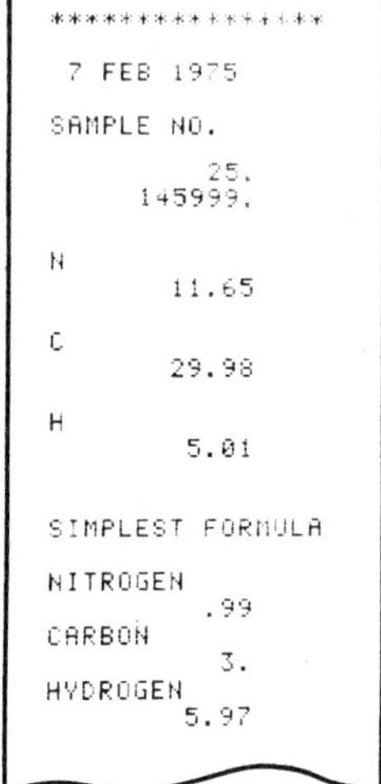

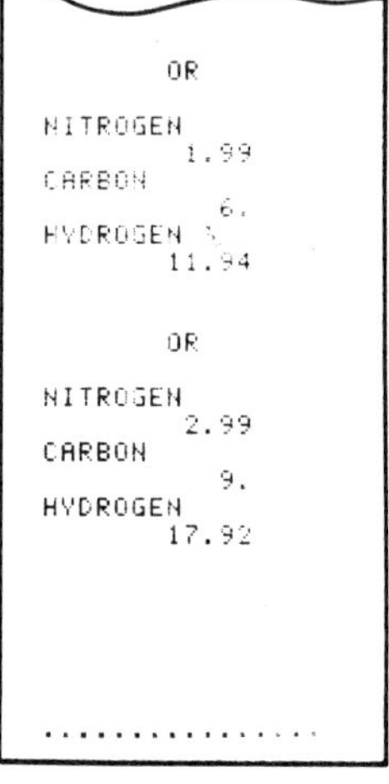

FIG. 20. Analytical report of Perkin–Elmer autosampler.

provide date, identification numbers, sample weight, percentages of each element and three empirical formulae.

3. *Data readout and processing*

When the Model 240 was introduced, a strip chart recorder was employed for presentation of the data in an easy to read form and its cost was relatively low. Although the recorder is still the most commonly used method of data presentation, it was inevitable that more sophisticated systems would be adapted. The range of methods currently employed will be briefly summarized here.

(a) *Recorder*. A recorder with an input of 1 mV is normally used and its corresponding accuracy in reading the signal is 0·25%; however this 1 mV input is not sufficient to record the range of signals normally sensed by the detectors and attenuation is needed to reduce the signal to a recordable level. A typical recorder trace is shown in Fig. 9 (p. 0).

(b) *DVM printers*. Signal attenuation is operator-dependent and has to be initiated by manual attenuators; moreover it reduces the final precision of the results. These drawbacks can be excluded by using analog-to-digital converters. A digital voltmeter combined with a conventional printer can be used for this purpose and an example of this system, as designed and used by the authors, is shown in Fig. 21. In addition the digital output of an electronic

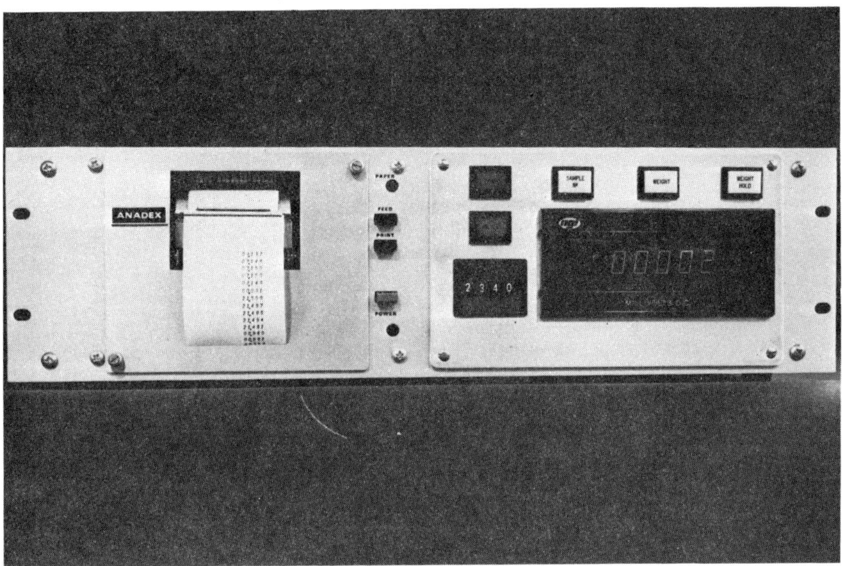

FIG. 21. Digital voltmeter and printer.

balance and the sample run number can be printed, resulting in a convenient data record of all relevant information. Many individual systems have been constructed and some are commercially available.

(c) *Programmable calculators and computers.* A natural development of the DVM-printer systems has been the addition of calculating circuitry and memory facilities, and the data-handling system described on p. 24 is an example, which is currently available from Perkin–Elmer.

The Model 240 can also be interfaced to large computers[11] for data storage and retrieval, calculations, as well as relation to other sample data. However, time-sharing of large computers has the drawback that the results may not be available on completion of the analysis, which could present problems if a repeat run of an unstable compound is required.

IV. PACKINGS AND FILLINGS

Various reagents, which are employed for combustion, reduction and absorption, form an integral part of the Model 240, and they require replacement on a regular basis. The manufacturers supply recommended reagents, but for special applications or individual preferences others may be required; these reagents have to be sufficiently stable to avoid damage to the instrument.

The Model 240 essentially reproduces the chemical techniques of Pregl and Dumas, therefore the instrument may be adapted to the modifications of the classical methods as described in the literature.

A. Combustion tube

The fillings for the combustion tube as shown in Fig. 22 were originally recommended by Perkin–Elmer. The positioning of the reagents in the

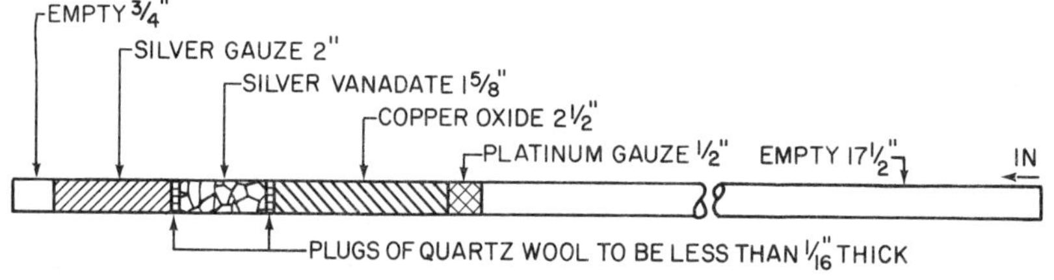

FIG. 22. Original Perkin–Elmer packings.

combustion tube is related to the temperature of the furnace at that point and this temperature is indicated between brackets.

Platinum gauze (950°C) is used for its catalytic action on certain compounds as well as for locating the remaining reagents in the tube, while the copper oxide (950°C) enhances the oxidation of the sample by providing oxygen. Silver vanadate (800°C) and silver gauze (500°C) absorb halogens, sulphur, some fluorine and some arsenic.

Copper oxide tends to fuse at 950°C and as a result its efficiency in the oxidation process decreases and it becomes difficult to remove from the tube. Macdonald and Turton[12] suggested a combination of copper oxide and magnesium oxide,as shown in Fig. 23, which was stated to be highly satisfactory for the regular analysis of fluorinated compounds.

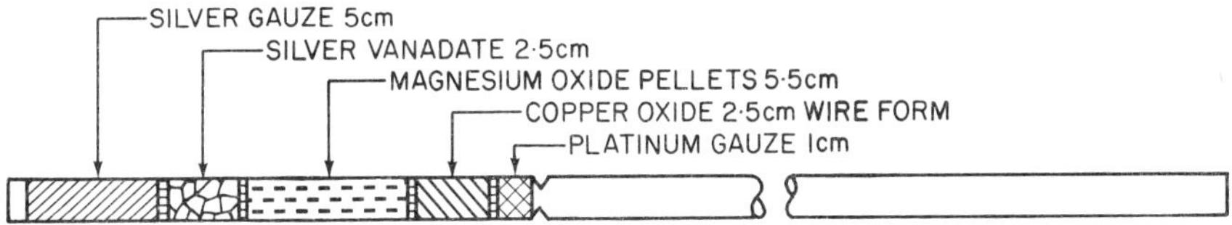

FIG. 23. Packings for combustion tube as suggested by Macdonald and Turton.

In 1966 Gustin and Tefft[13] recommended the use of predominantly silver-based reagents with the addition of cobalt oxide as is shown in Fig. 24. This proved particularly effective when a large number of halogen and sulphur-containing compounds were to be analysed.

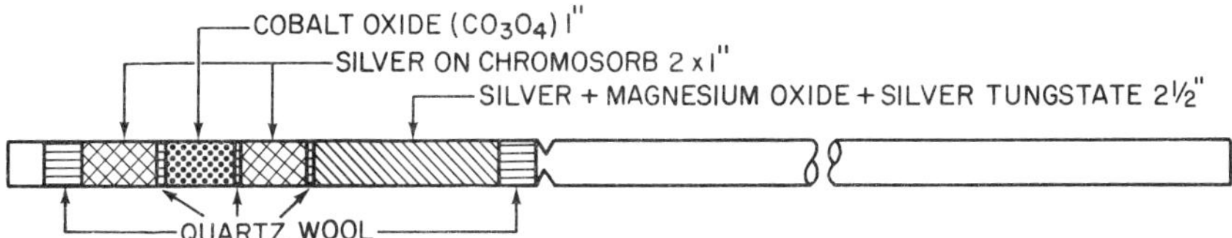

FIG. 24. Packings for combustion tube as suggested by Gustin and Tefft.

The work by Gustin and Tefft and recommendations in publications by Francis[14] and Yeh,[15] formed the basis of the packings shown in Fig. 25, which are at present recommended by Perkin–Elmer.[16] These packings are employed by the majority of Model 240 users.

Platinum gauze (950°C) is followed by magnesium oxide (950°C), which is coated by silver tungstate for catalytic purposes. Magnesium oxide can absorb a large quantity of fluorine; silver tungstate helps to maintain the magnesium oxide in pellet form. Silica forms the base for the next reagent

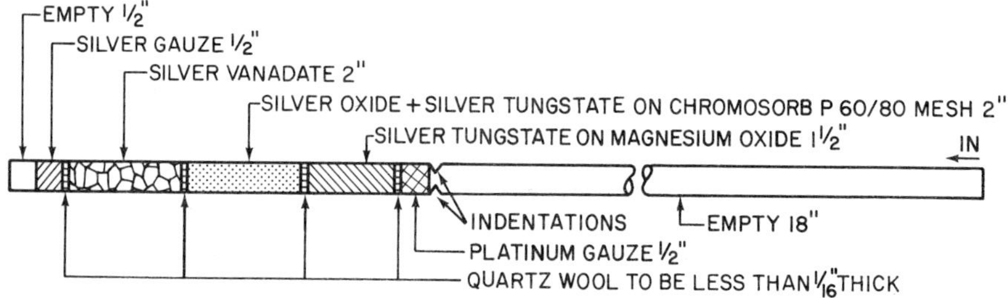

FIG. 25. Perkin–Elmer packings used at present.

containing silver oxide and silver tungstate, which enhance oxidation. Quartz tubes may devitrify and become porous on prolonged contact with this reagent, therefore regular inspection should be carried out. For absorption of halogens, silver vanadate is used; it is coated on to an inert base such as zirconium oxide to avoid powdering of the reagent. A roll of silver gauze completes the combustion tube packings and its major function is the removal of sulphur oxides. The reagents are separated by thin plugs of quartz wool.

The packings generally need replacing after 400 to 500 analysis, but this will depend on elemental composition and weight of compounds analyzed.

B. Reduction tube

The reduction tube contains copper granules, which remove residual oxygen from the combustion process and reduce the oxides of nitrogen to elemental nitrogen. Under normal operating conditions the copper is maintained at a temperature of 670°C, which provides the necessary conditions for reduction, but has less tendency to fuse the granules together as may happen at higher temperatures. It has been suggested that temperatures as low as 300°C can be tolerated in relation to the copper in the Model 240, but no data were found in support.

Two preparations of copper are commonly in use with the Model 240:

(1) wire form, which is supplied in 2 mm cut lengths,
(2) 60–100 mesh powder, which lasts about twice as long as the wire form.

Figure 26 shows details of the reduction tubes currently in use; the tube of smaller capacity provides about 150 analyses per filling and the other about 600 analyses per filling.

Because of the relatively high cost of the copper reagent, some users may

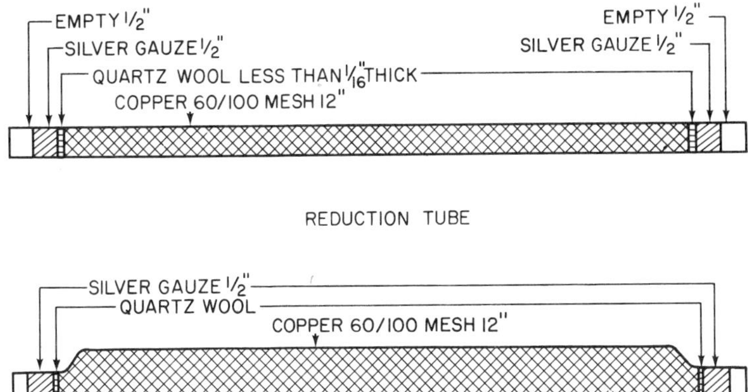

FIG. 26. Perkin–Elmer packings of normal and large diameter tube.

prefer to regenerate the reagent by reducing the oxidized copper retrieved from the instrument. Both hydrogen and carbon monoxide have been used.

The reduction by hydrogen, although quicker, results in copper which has a surface activity significantly lower than that of copper reduced by carbon monoxide, and moreover it is a rather dangerous procedure.

The disadvantages in the use of carbon monoxide are its toxicity and the length of time needed for the reduction.

Quartz wool plugs separate the copper from the silver gauze, which act as a secondary sulphur absorber in case a breakthrough should occur.

C. Absorption tubes

The four absorption tubes of the Model 240 can be divided into two pairs, namely "scrubbers", which remove water and carbon dioxide from the inlet gases oxygen and helium, and "traps", which absorb water and carbon dioxide formed during the analysis.

Figure 27 shows diagrammatically the two types of absorption tubes used in the Model 240.

Of the many reagents which are suitable for the absorption of water, magnesium perchlorate is recommended, for it has an excellent absorbing capacity for water. The most commonly used reagents for the absorption of carbon dioxide are sodium hydroxide and lithium hydroxide; the latter has a greater capacity for the absorption of carbon dioxide and in the preparation called "colorcarb" an indicator is incorporated.

A section of magnesium perchlorate is included in the carbon dioxide

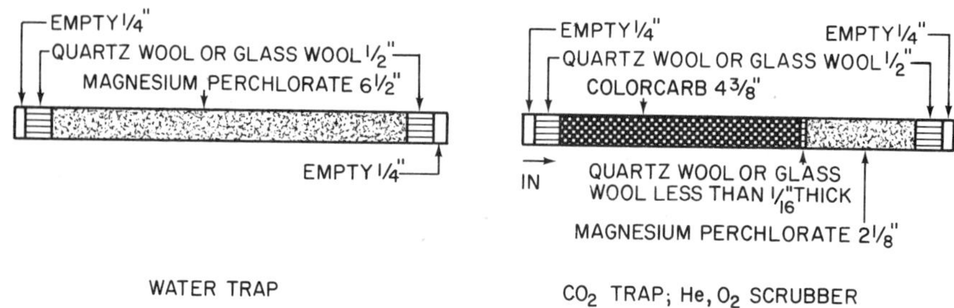

FIG. 27. Packings of the absorption tubes.

absorption tubes to absorb the water produced during the reaction between the alkali hydroxide and carbon dioxide.

Quartz or cotton wool plugs are placed at each end of the tubes and also between the water absorber and the carbon dioxide absorber.

D. Gas-purification tubes

A purification accessory can be attached to the gas inlet system before the scrubbers, in order to convert possible organic impurities to water and carbon dioxide and to remove traces of oxygen from the helium gas.

The purification tube for the helium gas is shown diagrammatically in Fig. 28. The tube contains copper granules to remove oxygen, platinum

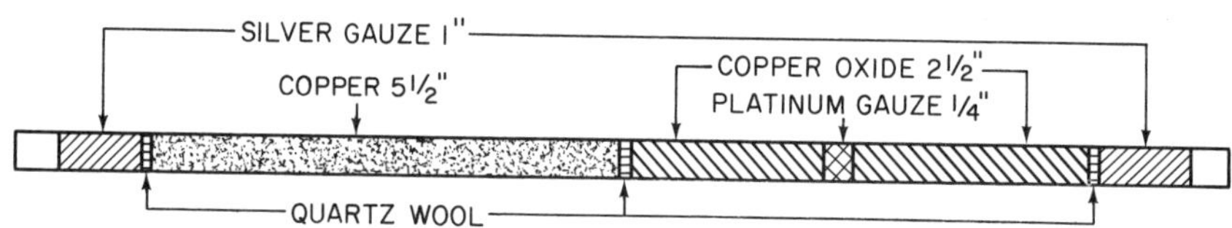

FIG. 28. Helium purification tube.

wool as catalyst and copper oxide for the oxidation of organic impurities. Silver gauze plugs are positioned at each end of the tube to locate the packings, as well as to absorb sulphur if present.

The diagram of the tube for the purification of the oxygen gas is shown in Fig. 29; it contains the same reagents as in the helium tube, but the copper granules are replaced by copper oxide.

Both purification tubes are maintained at approx. 625°C by a furnace unit (p. 22).

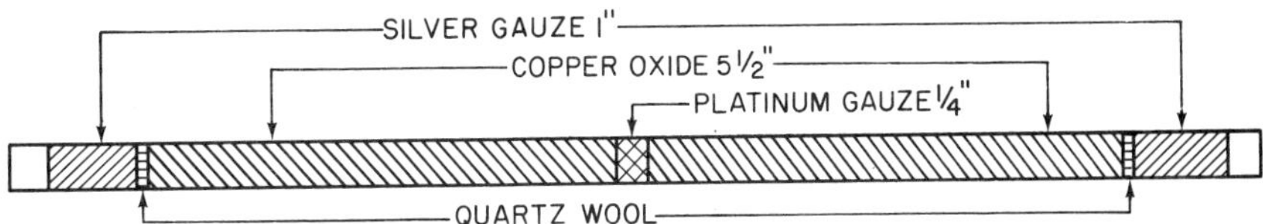

FIG. 29. Oxygen purification tube.

V. GENERAL PROCEDURE AND SAMPLE TREATMENT

Although the automation of the Model 240 has removed many of the manual operations which were necessary in the classical procedures, accurate preparation of instrument and samples remain of utmost importance in order to achieve accurate results.

Instrument and operator cannot always be blamed for all inaccurate results, since sample preparation starts outside the control of the microlaboratory. Assuming therefore that a pure sample has been submitted, the procedure described in this section should generally provide accurate analyses; slight variations in individual techniques may give equally good results.

A. Preparation of Model 240

When the Model 240 is used daily, it remains in the standby condition (p. 6), until an analysis is carried out.

It is recommended that when in the standby condition overnight, the furnace temperatures are reduced by 100°C and the pressure of helium is lowered to 12 p.s.i., in order to limit the operation of the pressure switch. Accordingly each morning the furnace temperatures and helium pressure must be restored to their normal operating levels.

1. *Conditioning of the instrument*

The constant purging of the instrument with helium has a drying effect on the chemicals and components, consequently conditioning must be carried out before any analyses are attempted. This conditioning is achieved by combusting one or two unweighed samples employing the same procedure as for a normal analysis, and it should be carried out if the instrument has not been used for an analysis for over an hour. As a result of this conditioning an equilibrium is created between the gases flowing through the instrument and those absorbed on to the surfaces of the chemicals and components. These conditioning runs can be utilized to adjust the setting of the helium inlet pressure to a level which allows the mixing volume to reach the required pressure of 1500 mm mercury in 60 to 90 seconds.

2. *Blank values*

Chemical absorbers remove most of the impurities from the gases before they enter the instrument but some (particularly nitrogen) will pass through and eventually reach the detectors. The quartz ladle, which is used for the transfer of samples to the combustion area, can also contribute to the blank value by retaining atmospheric gases on the surface, especially if devitrification has occurred in the quartz.

These effects are responsible for a small but significant signal at the detectors and they should be accounted for in the subsequent analyses.

Only one blank determination is usually required if environmental conditions in the laboratory are stable, and this should be carried out immediately after the conditioning run, by employing the normal analytical procedure but without a sample.

3. *Calibration of the instrument*

As the Model 240 detects only part of the gases from the combustion of a sample, a calibration constant has to be determined for each detector bridge. This constant relates the number of microvolts detected to each microgram of element. Typical values are 8 μV/μg for nitrogen, 22 μV/μg for carbon and 65 μV/μg for hydrogen.

A wide range of pure organic compounds can be used as standards[6] to determine these constants and some are listed in Table 1.

Table 1. Some organic compounds used as standards for calibration

Name of standard	Percentages of elements		
	%N	%C	%H
Phenacetin	7·82	67·02	7·31
Acetanilide	10·36	71·09	6.71
2,4-Dinitrophenylhydrazine	28·28	36·37	3·05
Melamine	66·64	28·57	4·79
Anthracene	—	94·34	5·66

Standards such as acetanilide or phenacetin generally result in the required accuracy for most compounds.

In order to achieve the highest possible accuracy, however, it is recommended to choose a standard compound which is similar in structure and composition to the sample to be analyzed (p. 00). Although it is impracticable to take this recommendation as a rule when continuously analyzing samples, it should be considered if difficulties are encountered in meeting the desired accuracies.

Calibration of the instrument is generally performed twice daily, but it may be necessary to calibrate more frequently depending on the variety in the composition of the samples and/or changes affecting the performance of the instrument; the procedure is the same as for a normal analysis.

B. Preparation of the samples

Stable compounds provide little difficulty at the weighing stage since the popular and convenient platinum boat can be employed. However, certain compounds or additives could attack the platinum and consequently a porcelain or aluminium boat may be required.

Details of the common implements for microanalytical work, such as platinum boat, platinum-tipped forceps, cold block, spatula, weighing pig, etc., are described elsewhere.(6)

For unstable compounds several sample containers can be employed and they will be described separately.

1. *Stable compounds*

A platinum boat can be cleaned by washing it in acid or melting a borax brazing flux in it, which will remove most contaminants. After washing with water, the boat is heated to redness and by means of platinum-tipped forceps transferred to a cooling block, where it is brushed clean of dust particles.

The boat is then placed on the balance pan and on most balances its weight can either be recorded directly or it can be counterpoised. Electronic balances, however, have a limited display of digits and counterpoising the weight of the boat enables its most sensitive range to be utilized. Some electronic balances require regular calibration to compensate for changes due to varying load, by employing accurate standard weights.

The platinum boat is transferred to the cold block where about 2 mg of sample is placed in the boat; after brushing of its base, the boat is returned to the balance pan, and the exact weight of the sample is recorded.

At this weighing stage careful attention should be paid to loss or gain of weight, which can be due to either solvent evaporation, sample volatilization, or water uptake by hygroscopic samples. When a loss of weight is noticed, the sample should be further treated in order to remove the solvent and special treatment or weighing techniques can usually overcome most problems in connection with volatile or hygroscopic compounds (p. 41).

Some samples may require an oxidant and/or a special absorber (p. 43), which can be added to the boat when it is returned to the cold block.

The quartz ladle is removed from the instrument and the boat transferred from the cold block to the open end of the ladle (p. 37), which is then returned to the cold section of the combustion tube. Once the tube has been sealed, the program sequence of the instrument can be initiated by push button and the analysis will be taken to completion as described previously (p. 6).

After the analysis has been completed, the boat can be reweighed to determine a possible residue if no additives had been employed.

In the automatic sample introduction system (p. 23) a different type of sample container is employed, which is shown in Fig. 30. The sample is

FIG. 30. Sample containers for autosampler shown with ordinary platinum boat.

placed in the container between two quartz wool plugs; several containers are placed into the magazine and the subsequent introduction and analysis of the samples is carried out automatically by the instrument.

2. *Unstable compounds*

Several methods exist for weighing of unstable compounds, but the most widely used method for analysis in the Model 240 is enclosing the sample in an aluminium capsule. However, some samples may react with the aluminium and alternative materials such as glass or indium have to be used.

(a) *Aluminium capsule*. Perkin–Elmer manufacture aluminium capsules[17] and an encapsulating tool, which are shown in Fig. 31.

The top and bottom sections of the capsule are cleaned in acetone and air-dried. They are then heated in a furnace at 500°C and stored in a desiccator. After the two halves have been brushed and weighed together, they

FIG. 31. Perkin–Elmer encapsulating tool and cold block.

are transferred to the encapsulating tool where a sample is introduced in the bottom section by means of a spatula or syringe. The top section is placed over the bottom one and by manually bringing the dies of the tool together an hermetic seal is created around the edge of the capsule. After sealing, the capsule is returned to the balance for reweighing and subsequently placed in the ladle for introduction into the instrument.

Highly volatile compounds can be successfully analyzed by this encapsulating technique.

The causes of incorrect analysis by this method can usually be related to creeping of samples such as oils, enclosure of air inside the capsule, or reactivity of the aluminium at the combustion temperature.

To overcome the variation due to creeping of the sample outside the seal during encapsulation, the capsule can be washed in acetone, air-dried and wiped with chamois leather before the weight of its sample is established.

Enclosure of air inside the capsule causes a significant increase in the nitrogen result and a blank determination using an empty capsule may help. The degree of volatility of the sample can cause varying amounts of air to be enclosed, therefore sealing under helium or oxygen may become necessary (p. 40).

An important danger in the use of aluminium capsules is the reactivity of the aluminium on contact with the silica from the combustion tube. Therefore, the aluminium capsule should never be allowed to stay behind in the combustion tube and preferably a ladle made from a material other than quartz should be used (p. 37).

Under certain combustion conditions a large proportion of the available oxygen in the combustion area can be used for the formation of aluminium oxide. The resulting depletion of oxygen can be prevented by the addition of an oxidant to the sample in the capsule.

(b) *Capillary*. Liquid samples can also be enclosed in a glass capillary, which is made from a conventional 1 mm melting point tube, as shown in Fig. 32.

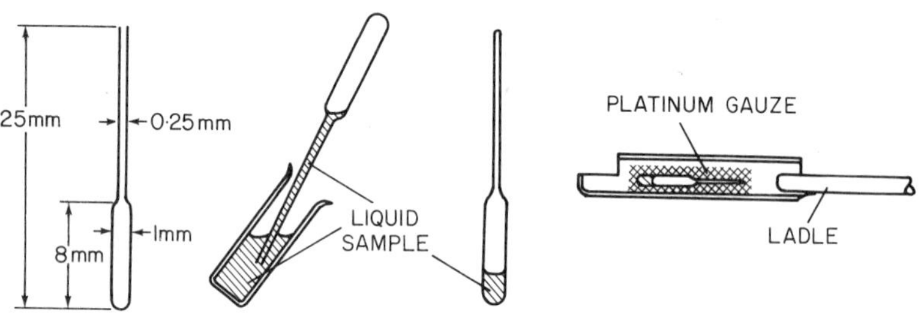

FIG. 32. Capillary method for liquid samples.

The empty capillary is weighed and after lightly heating the sealed end to expel some of the air, the open end is dipped into the liquid sample, which is drawn up into the tube on cooling of the glass. The liquid is then centrifuged into the sealed end and the fine capillary is sealed by lightly heating it. After reweighing the sealed tube to establish the sample weight, it is placed in a roll of fine mesh platinum gauze and inserted into the conventional quartz ladle to be analysed in the normal way.

The problem of enclosed air in the capillary is similar to that of the aluminium capsule, but it is more difficult to overcome since sealing by heat in helium or oxygen is required.

A further problem is the difficulty of removing the glass fragments from the gauze, ladle or combustion tube on completion of the analysis.

(c) *Indian tube*. An alternative to the glass capillary is a capillary made from indium,[18] which has a low melting point and therefore can be sealed easily. However, indium is costly and it can leave objectionable residues behind in the ladle and the combustion tube.

3. *Ladles for transfer of the sample*

A quartz ladle as shown in Fig. 33 is usually employed for the transfer of

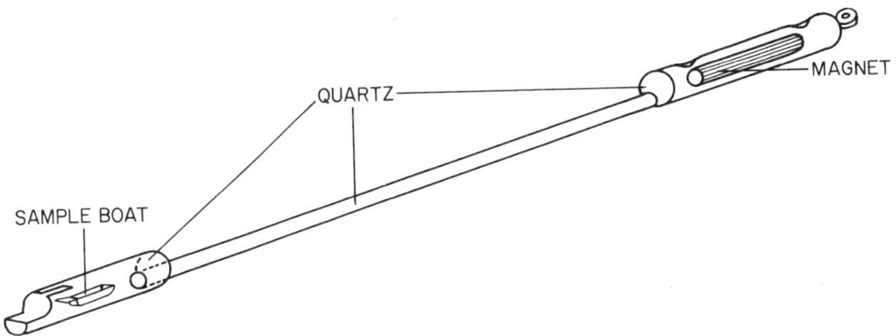

Fig. 33. Normal quartz ladle.

the sample from the cold section to the hot area of the combustion tube. The sample in its container is placed in the open end of the ladle and the ladle subsequently placed in the cool part of the combustion tube, which is then sealed by means of the entrance plug. It is recommended to wear gloves when handling the ladle, in order to avoid the introduction of impurities.

The automatic sequence of the instrument is started by a push button on the front panel and after a few minutes the inject light will indicate that the system is ready for introduction of the sample into the combustion furnace. This introduction is accomplished by means of a hand-held magnet and the automatic sequence of the instrument is continued by pushing the combust button.

When a sample is weighed in an aluminium capsule, it is advisable to use a ladle made of a material other than quartz, since aluminium has a deleterious effect on quartz. For this purpose the quartz ladle can be modified be replacing the sample carrier by a platinum sheet cylinder, or stainless steel cup. Figure 34 shows a modified ladle as designed by the authors, in which the quartz sample carrier is replaced by a low carbon stainless steel cup.

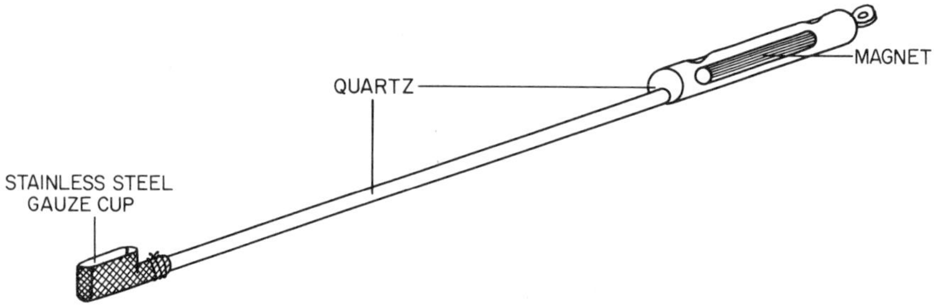

FIG. 34. Modified ladle for aluminium capsules.

C. Combustion procedure

The automatic combustion cycle of the Model 240 and the recommended combustion tube fillings (p. 27) will generally produce satisfactory results for the majority of compounds analyzed. Troublesome compounds will be discussed separately, together with various methods of improving the combustion conditions (p. 42).

D. Data processing

Unless a computer-based data handling system is used in conjunction with the Model 240, some transfer of data and calculations have to be carried out manually. The recorded data are usually transferred to a form on which all data relevant to a particular analysis are collected.

Figure 35 shows a calibration data sheet as designed by Perkin–Elmer and as an example the calculations for a calibration run of acetanilide are filled in.

The form shown in Fig. 36 is of a simpler design and shows only the basic information in connection with the analysis of a sample. Of the many different means of calculating the percentages, the programmable calculator is becoming increasingly more popular, since it removes the simple but tedious calculations involved.

VI. SPECIAL TREATMENT OF TROUBLESOME COMPOUNDS

The majority of organic compounds analysed daily in Universities, Colleges, and Industries such as Pharmaceutical, Polymer, Coal, Petroleum, etc., give satisfactory results when the general procedure of analysis for the Model 240 (p. 31) is followed.

The occasional troublesome sample may need some special treatment

Calibration Data Sheet

A. Standard Name: ACETANILIDE Date: 2.3.76

C. Standard Weight: 1956 µg

D. Theoretical percentages 10·36 % N; 71·09 % C; 6·71 % H

E. Theoretical weights: 202·64 µg N; 1390·52 µg C; 131·25 µg H

F. Total signals:

Total signal (in µV) = Read + C Suppression* - Blank Value - Zero where:

Read or Zero (in µV) = (Recorder Reading X10) (Attenuation)

Blank Value (in µV) = Blank Read - Blank Zero

	N Signal	C Signal	H Signal
Reads (µV)	465 X 4 = 1860	75 X10 = 750	554 X 16 = 8864
+ C Suppression* (µV)	***	Setting 3 = 30,000	***
- Blank Values (µV)	= 75	= 90	= 220
- Zeros (µV)	160 X 1 = 160	12 X10 = 120	80 X 1 = 80
Total signals (µV)	= 1625	= 30540	= 8564

G. Sensitivities:

$$K = \frac{\text{Total signal } (\mu V)}{\text{Theoretical weight } (\mu g)}$$

K_N 8·02 $\frac{\mu V}{\mu g}$; K_C 21·96 $\frac{\mu V}{\mu g}$; K_H 65·25 $\frac{\mu V}{\mu g}$

Values of C Suppression for C ATTENUATION Switch Settings. (Use in Computation of C Signal only).

Switch Setting	0	1	2	3	4	5
C Suppression (µV)	0	10,000	20,000	30,000	40,000	50,000

FIG. 35. Perkin–Elmer calibration data sheet.

SAMPLE DATA SHEET

DATE ANALYSED :	2.3.76	M.A. DEPT. No :	1234
DATE SUBMITTED :	2.3.76	SUBMITTED No :	579–26
FORMULA :	$C_{10}H_{13}NO_2$	WEIGHT :	2 100 μg

		N	C	H
SIGNALS	READ :	384	1150	649
	Attn./Supp :	4	30,000	16
	TOTAL :	1536	31150	10384
	ZERO :	155	120	85
	BLANK :	75	90	220
	TOTAL	1306	30940	10079
SENSITIVITIES	μV/μg :	8.02	21.96	65.25
PERCENTAGE FOUND	:	7.75	67.09	7.36
CALC.	:	7.82	67.02	7.31
RESIDUE	:			
MOISTURE CONTENT	:			

REMARKS:-

FIG. 36. A simplified sample data sheet.

during its preparation or its combustion, and the various methods to overcome the difficulties will be discussed here.

A. Environment-sensitive samples

1. *Air-sensitive samples*

Several methods can be employed to eliminate environmental effects, and the choice depends on the sensitivity of the sample. In most cases the sample has to be weighed and sealed in a container, preferably under inert conditions, in order to avoid the environmental effects until the sample is introduced into the Model 240.

Various sample containers can be sealed and they have been discussed together with their problems (p. 43). The most versatile and widely used is the aluminium capsule, which is supplied together with an encapsulating tool by Perkin–Elmer.

A simple and effective method of removing air from the aluminium capsule is to replace the air by oxygen or helium during sealing, by placing the outlet of a tube connected to an oxygen or helium supply near the capsule. The helium supply of the Model 240 in standby condition can be used for this purpose, by replacing the normal entrance plug by a similar plug with an exit tube (a plug of this description is supplied with the instrument for checking the oxygen flow). However, it should be remembered that the entrance port to the instrument should not be left open for longer than one minute, since the pressure may drop to a level which causes the detector safety circuit to operate.

In severe cases a glove-box arrangement filled with helium can be coupled directly to the combustion tube inlet of the Model 240, or used on its own. The size of the glove-box should allow several items to be placed inside, such as balance weighing compartment, cold block, sample encapsulating tool, capsules, etc. Fig. 37 shows a glove-box arrangement connected to the Model 240.

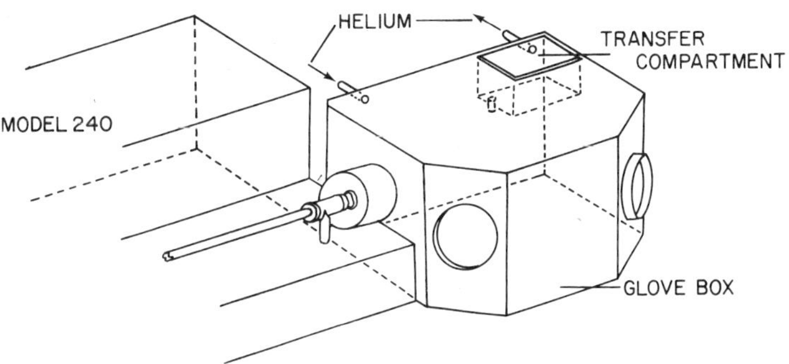

FIG. 37. Glove-box connected to Model 240.

2. *Volatile samples*

Weighing the sample in aluminium capsules is the most successful method of handling volatile samples. Depending on the degree of volatility of the sample, varying amounts of air can be enclosed with the sample in the capsule (p. 36) and this should be prevented by one of the methods described above, otherwise high nitrogen results will be obtained.

It is sometimes difficult to judge the amount of sample needed for an analysis particularly if a liquid of low boiling point is handled and it may be helpful to use a syringe; however, trial and error is usually the only way.

3. *Hygroscopic samples*

These samples can be analyzed successfully when they are sealed in aluminium capsules, in the same way as for volatile samples. If the preparation of the sample is not carried out in a helium atmosphere i.e. in the above-mentioned glove box system, every effort should be made to handle the sample in the shortest possible time to reduce water-uptake to a minimum.

The use of a weighing pig has been described in the literature,[(6)] but it has no advantage over the use of aluminium capsules or the following timed weighing procedure in an open boat, which is preferred by the authors.

After weighing a platinum boat, the sample bottle is opened and a stop-watch is started; the sample is then transferred to the boat as quickly as possible. When a modern electrobalance is used, which stabilizes very quickly, a weight can usually be recorded in 30–60 seconds from opening the sample bottle. Further weights should be recorded every 15 seconds and when sufficient data are available they can be extrapolated back to the time the bottle was opened, as well as extrapolated in the opposite direction to allow for further uptake of water in the time of transfer from balance to the instrument. The carbon and nitrogen results can be related to the sample weight at zero time, while the hydrogen result will represent the total water from both the original sample and absorbed water; the latter will therefore have to be accounted for in the final hydrogen calculation. It should be noted, however, that in many cases this small amount of absorbed water will usually be removed by the dry helium flowing through the instrument and purged to atmosphere before the combustion of the sample takes place.

It may be preferred to allow the sample to become saturated before it is transferred to the Model 240 and then relate the instrumental results to a separate determination for the water, such as a Karl Fischer. This method is rather time-consuming and moreover after introduction of the sample into the Model 240, part of this water will likely be removed by the flow of dry helium as described above. This drying process may be assisted by the slightly elevated temperature in the combustion tube and consequently the final results may be misleading. The temperature of the combustion tube near the flash heater can be reduced by placing a cold wet filter over the particular area or blowing cold air over it.

It may be helpful to remember that the dry conditions in the Model 240 can be utilized to remove free water or solvent from a sample when the instrument is in standby condition. However, since several manipulations of the sample boat are involved, care should be taken to avoid spillage of the sample.

4. *Light-sensitive samples*

Aluminium capsules are suitable containers for light-sensitive compounds, and speed in sample transfer or working in light of an acceptable wavelength or intensity can minimize these effects.

B. Various methods of aiding the combustion

The automatic combustion cycle of the Model 240 will not always succeed in combusting all samples; certain compounds may prove difficult and this can usually be overcome by manipulating some factors concerning the combustion:

1. *Improved use of available oxygen*
The sample can be brought into contact with fresh oxygen by moving the ladle a few times inside the combustion furnace, which disperses the combustion products.

2. *Prolonged time of combustion*
The temperature of the combustion furnace can be maintained for extra time by turning a switch on the front panel to the "hold" position 30 seconds after the combust button has been pressed, but it should be remembered that due to backflushing of helium in the mixing volume the gaseous equilibrium may be affected by extensive prolonging of the "hold" position; this effect can be reduced by switching to the "hold" position 60 seconds after the combust button has been pressed, at which time helium has stopped back-flushing the mixing volume and therefore the gaseous equilibrium is maintained."

3. *Increased temperature of the combustion furnace*
The temperature of the combustion furnace can be increased to about 1000°C by a control on the front panel of the instrument.

4. *Alternative use of the flash heater*
The samples can be withdrawn to the centre of the flash heater, which also reaches a temperature of about 1000°C and in addition this temperature can be maintained for any length of time by turning to the "hold" position as soon as the flash heater switches on.

5. *Addition of donors*
Oxygen donors,[19,16] such as cobaltic oxide, tungstic oxide, or vanadium pentoxide, can be added to the sample in its container, or an additive such as powdered tin, which increases the temperature in the immediate vicinity of the sample. A ceramic boat should be employed when tin is used, since platinum is attacked at the resulting higher temperature of combustion.

Donors should be specially prepared in order to avoid variable blanks; for example the most commonly used tungstic oxide is heated at 500°C and stored in a desiccator prior to use. When donors are used, sample-derived residue determinations are usually prevented and continuous use of donors has a deleterious effect on the quartz of the ladle and the combustion tube, due to their spattering effect.

6. *Varying the packings of the combustion tube*
Some variations of tube packings have been discussed (p. 26 and p. 44), but they should only be considered if certain types of compounds are to be

analyzed regularly. The recommended tube packings satisfy generally the majority of compounds.

It will be apparent from the variations described above, that the combustion system of the Model 240 can be adapted quite readily to suit virtually all types of compounds.

C. Problems related to type of compound

1. *Nitro-compounds*

Most heterocyclic nitrogen compounds can be analyzed without difficulty, but some poly-nitro derivatives may give low nitrogen results, which according to Swift[20] is probably due to absorption of some nitrogen oxides by the copper oxide formed in the reduction tube. Lower operating temperature of the reduction furnace (580–600°C) or the use of aluminium containers were reported[21] to provide acceptable results, while the authors found that in certain cases the addition of vanadium pentoxide to the sample was beneficial.

2. *Halogenated and sulphonated compounds*

Most compounds containing the common elements halogens and sulphur can be analyzed successfully in the Model 240, since the reagents employed in the recommended packings (p. 27) have a large capacity for the removal of interfering combustion products.

If samples containing predominantly halogens or sulphur are analyzed regularly, it may be useful to extend certain packings in the combustion tube, i.e. magnesium oxide for fluorine[12] and silver reagents dispersed on an inert support for halogens and sulphur,[13] in order to increase the capacity for absorption of these specific elements, and consequently the lifetime of the packings will be extended.

Analyzing many fluorine compounds can result in devitrification of the quartz tube, and it can partly be prevented by the addition of magnesium oxide to the sample in its container, which absorbs the offensive fluorine.

Some highly sulphonated compounds can cause a yellowish film on the tube to the right of the flash heater. This deposit does not usually affect the results and it can be burned off in an atmosphere of oxygen.

3. *Organo-silicon and organo-phosphorus compounds*

The presence of silicon or phosphorus in the samples can cause problems; stable refractory complexes with carbon, such as silicon carbide, may be formed, or carbon may be occluded by residues in the boat. These problems can be overcome by mixing the sample with either vanadium pentoxide or tungstic oxide[19] or alternatively a combustion temperature of 900°C and a combustion tube packing of mainly magnesium oxide.[22] The authors found that extra combustion time was often required with silicon compounds and generally few problems were encountered with compounds containing phosphorus.

4. *Organo-metallic compounds*

Organo-metallic compounds can lead to the formation of metal carbides and metal carbonates, which can often be prevented by the addition of tungstic oxide or vanadium pentoxide,[21, 23] or alternatively by varying the packing of the combustion tube.[15]

Some volatile metals may poison the tube packings, therefore samples containing such metals as arsenic, selenium, antimony, bismuth and mercury, should be analyzed when the combustion tube packings approach the need for changing. The addition of gold wool at the end of the combustion tube may be beneficial when analyzing mercury compounds.

The amount of metal present in the sample can often be determined from the ash residue, unless volatile complexes such as metal chelates are handled, or if additives were used in the sample preparation.

5. *Inorganic compounds*

Many inorganic compounds can be analyzed without modification to the normal analysis cycle of the instrument. The formation of carbonates can be overcome by the addition of extra oxygen from a donor such as tungstic oxide or vanadium pentoxide.

6. *Boron compounds*

The addition of powdered tin is recommended when boron compounds are analyzed, since the possible formation of a carbide cannot always be prevented by the addition of an oxidant.

7. *Carbon fibres and graphite*

The Model 240 supplies sufficient oxygen for the combustion of these compounds, but a prolonged combustion and higher combustion temperature as described previously, is nearly always required.

D. Special applications

1. *Organic matter in water or atmosphere*

The Model 240 has been successfully applied to study the organic contents of waters, such as rivers, lakes[24] and seas.[25, 26] A known volume of water is filtered through a metal[25, 26] or glass fibre filter,[24] which is subsequently analyzed in the instrument, usually only for carbon and nitrogen.

In similar studies carried out by the authors, it was found that the glass fibre filters melted in the ladle at the combustion temperature, and this fusion of glass and quartz could be prevented by the use of a special ladle made from stainless steel. Care must be taken in the preparation of the filter discs, since high and variable blanks may occur.

An alternative method to filtering of organic matter from the water, is the direct analysis of water samples in the Model 240. The large amounts of water released during the combustion can be removed by a water absorber located between the outlet of the combustion tube and the inlet of the reduction tube, after which the small amounts of carbon and nitrogen can be detected.

A similar application is the determination of carbon, hydrogen and nitrogen in atmospheric aerosols,[(27)] whereby the organic matter was collected on glass fibre filters, part of which were subsequently wrapped in platinum gauze and placed in the quartz ladle for the usual analysis.

2. *Organic matter in rocks, shales and soils*

As low levels of the elements carbon, hydrogen and nitrogen are found in these samples, a comparatively large amount of sample has to be employed in order to obtain a reasonable signal at the detectors and consequently a larger size boat is required.

3. *Petroleum compounds*

Aluminium capsules are useful sample containers for oils, but sometimes the creeping effect of these compounds can cause variable results and a method of overcoming this problem has been described previously (p. 36).

For the determination of carbon, hydrogen and nitrogen in petroleum compounds, the normal procedure for the Model 240 can be followed. Sometimes low levels of nitrogen (less than 0·1%) have to be determined and it is recommended that the sample weight is increased to 5–10 mg in order to achieve a detectable signal for nitrogen; the resulting carbon and hydrogen signals should be ignored, since they may well fall outside the linearity range of the detectors (p. 48).

In an evaluation study by Smith *et al.*,[(28)] typical petroleum products were analyzed in the Model 240 for carbon, hydrogen, nitrogen and oxygen, and the results compared with carbon and hydrogen analysis by the Pregl method, oxygen by the Unterzaucher method, and nitrogen by the automated micro Dumas analyzer. The results obtained by the automated method were found comparable to those obtained by conventional methods, while for the analysis of nitrogen the Model 240 was preferred, since levels of nitrogen from 0·05 to 0·1% present difficulty for conventional methods.

VII. ACCURACY AND SOURCES OF ERROR

A. Accuracy of the results

The results of an elemental analysis for carbon, hydrogen and nitrogen have traditionally been regarded as acceptable, if the accuracy of the results is within 0·3% absolute of the theoretical value.

The Model 240 is specified to achieve results within 0·3%, but in practice the majority of results will be much closer to the theoretical value than this.

It has been shown by Wagner[(29)] at a meeting of the Microchemical Society at the Pennsylvania State University in 1968, that the absolute accuracy of the Model 240, as compared to the classical methods, is generally superior. Wagner presented the results of a statistical evaluation based upon hundreds of individual determinations for each method and each element using a range of standards as well as pharmaceutical compounds, and concluded that the accuracy of the Model 240 was equal to the classical techniques for hydrogen and substantially better for carbon and nitrogen.

Another comparison between the Model 240 and the classical procedures was made by Smith *et al.*,[(28)] who determined carbon, hydrogen, nitrogen and oxygen in petroleum compounds; the results for carbon and hydrogen were comparable, while the Model 240 was preferred for the low level determination of nitrogen in petroleum samples since this presented difficulty for conventional methods.

The results for nitrogen in highly fluorinated samples by the Model 240 are comparable to those obtained by the Kirsten method, which has been used for such compounds, as the Dumas method is quite unsatisfactory.

The Elemental Analyzer User Forum of the Microchemical Methods Group carried out a collaborative study of C, H and N elemental analyzers,[(30)] based on a series of results obtained by fourteen Model 240 users and one Carlo Erba user on the same sample (3-chlorophenothiazine). This study was mainly aimed at comparing results routinely obtained by several elemental analyzers of varying age and in different laboratories on the same sample, which had been accurately analyzed in one laboratory by classical procedures as a reference. The precision of the results obtained by the elemental analyzers was generally lower than those obtained by the conventional methods, but factors such as maintenance and operator skill could obviously not be quantified.

Although the automation of the Model 240 has removed the need for many of the manual skills involved in the classical methods, the technical skill of an operator can be an asset for locating the causes of inaccuracy, and ultimately the excellent results the Model 240 is capable of, can be achieved.

Parameters which can influence the accuracy of the results will be discussed separately, in order to assist in locating possible causes of error.

B. Sources of error

1. *Environmental conditions*

The temperature and humidity of the environment in which the Model 240 is placed, should ideally be kept constant, although small changes will not normally affect the performance of the instrument.

Large and sudden changes in temperature, however, can influence the reproducibility of the instrument and of the regulators used for helium and oxygen supplies, while changes in humidity can affect the hydrogen results, due to varying blanks.

In addition the performance of the balance may suffer when variations in temperature or humidity occur, and ideally a balance should be placed on an anti-vibration table in draught-free surroundings.

2. *Weighing and sample handling*

A sample weight of 1 to 3 mg is generally recommended for the analysis of organic samples, and it should be weighed to an accuracy of at least 0·1%. Consequently, the balance must be capable of achieving this accuracy and the analyst must have developed a weighing technique matching these requirements.

Modern electrobalances can obtain an accuracy as low as 0·01% related to a 1 mg sample, but they should be checked regularly for zero stability and consistency of calibration, in order to avoid a gradual and often unnoticed deterioration in performance.

3. *Instrumental control of pressure and temperature*

Constant pressure in the Model 240 is an important parameter for good performance; inaccurate pressure reproducibility can be caused by the regulation at the helium inlet stage, leaks or restrictions in the system, or malfunction of the pressure switch.

Large changes in environmental temperature can affect the temperature stability of the oven unit, which temperature controls the pressure switch and other sensitive components, and it can become apparent by a gradual change of the calibration factors during the day.

4. *Detectors*

Two potential sources of error are present in the detectors, and these are a mole-fraction effect and a non-linearity effect. The mole-fraction effect is caused by changes in concentration of the gases before and after absorption of water and carbon dioxide; this effect will cause a small error in the read-out of a thermal conductivity detector, because the sample side will be referenced against a gas mixture of different mole fraction. The non-linearity effect will only become apparent if very large signals are generated, which can fall outside the linearity range of the detectors.

Both these errors will not normally become significant, unless sample weights of larger than 5 mg are employed or compounds with wide variations in elemental concentration are analyzed; the latter can be compensated for by calibrating the instrument with a standard of a composition similar to that of the sample.

The output of the battery used in the fixed attenuation circuit of the carbon bridge, will gradually reduce and monthly checking of the fixed attenuation voltage against the 1 mV recorder read-out can prevent the resulting small error.

5. *Combustion, reduction, and absorption tubes*

The condition of the quartz tubes and all packings can have a marked effect on the quality of the analytical results.

The quartz of the combustion tube will devitrify at the operating temperature and it should be checked regularly and replaced if porosity is suspected, which can manifest itself by high nitrogen results. The reduction tube will be similarly affected, but it will take longer to appear since its operating temperature is lower.

The efficiency of the packing materials of the combustion tube deteriorates over a period of time when they remain at their normal operating temperatures and regular replacement of the packings can avoid the resulting error.

Elements such as phosphorus and sodium can poison the packings, causing errors to following analyses; this poisoning effect can be reduced by the addition of tungstic oxide to the sample.

When the tubes are repacked, care should be taken to avoid too densely or too loosely packed reagents, which can cause respectively restriction to the gas flow and therefore affect the pressure, or on the other hand cause breakthrough of combustion products due to channelling. Both conditions can cause serious errors in the determination of the elements.

6. *Recorder readout*

The recorder used with the Model 240 has normally an input of 1 mV full scale deflection and a quoted precision of 0·25%. Consequently, it is unlikely that the readability error will be less than 2·5 μV and, if a signal attenuation factor is applied, the error will be 2·5 μV times the attenuation factor.

Since on average the carbon signals are usually much higher than either the hydrogen or nitrogen signal, a fixed suppression is applied to the carbon bridge by means of a battery circuit, which enables high signal levels of carbon to be recorded with little loss of accuracy.

The readability of the nitrogen signal, however, can give errors which are dependent on the concentration of nitrogen in the sample analysed. For example, when two samples of 2 mg weight each, respectively containing 5% and 50% nitrogen, are analyzed, the corresponding errors due to the 0·25% readability accuracy will be 0·02% and 0·25%. In order to improve the readability of the nitrogen signal when compounds of high nitrogen content are to be analyzed, a sample of a lower weight should be employed.

Small errors can be caused by the recorder itself if it is not maintained in good order (p. 54). In addition the printing from one chart to another may slightly differ, which could be checked by means of a standard 1 mV input source and any necessary adjustments made to the span of the recorder.

7. *Digital readout devices*

A significant improvement in accuracy can be obtained by replacing the recorder with a digital printer or computer, which will remove the readability error of the recorder and the additional increase of this error due to attenuation factors. This is particularly relevant to high nitrogen signals as discussed previously.

The input range of a digital printer or computer can accommodate most signal levels generated for the three elements without the need for manual signal attenuation and their readability error is generally only 1 μV. The inbuilt suppression on the carbon bridge can also be removed, but this will only result in a slight improvement in the accuracy for this element; it does however remove the need for monthly checking of the battery output (p. 49).

VIII. MAINTENANCE

The latest Perkin–Elmer Manual[7] supplies extensive information on maintenance and trouble-shooting on the Model 240, therefore the present section is mainly aimed at providing additional information which may be of assistance in maintaining the instrument in good working order.

The routine maintenance on the Model 240 comprises replacement of packings, tubes and gases, cleaning of solenoid valves, and adjustment to carbon detector output and recorder.

The present discussion will be limited to replacement of packings and tubes, checking of recorder and finally some modifications of positioning the programmer-wheel to the side of the instrument will be discussed.

The most economical time for maintenance of the instrument is at the end of the week, since the instrument can subsequently stabilize itself over the weekend. Regular replacement of packing materials, rather than waiting for a decrease in performance, can avoid loss of analyses.

It is important to remember to switch off the power supply to the expensive detector block during maintenance, in order to prevent possible damage to the detector filaments due to overheating. This will apply especially when the gas flow through the detectors will be interrupted.

A. Replacement of the packings

The frequency of replacing the packings will depend upon factors such as the number and types of compound analyzed, the amounts of sample used, the amount of oxygen introduced during each analysis, the number of blank determinations performed and the quality of the gases employed.

Typical usage of the various packings and gases are given in Table 2.

Table 2

Packings or gases	Average number of analyses
Tube packings:	
Combustion	400–500
Reduction, normal	150–200
Reduction, large capacity	600
Scrubber, helium	800
Scrubber, oxygen	1800
Traps	150–200
Gas cylinder:	
Helium (260 ft^2)	800
Oxygen (50 litre)	1800

1. *Combustion and reduction*

The packings of the combustion tube will gradually breakdown when they are kept at the normal operating temperature, even when analyses are not performed. In addition the temperatures of the combustion and reduction furnaces can increase overnight, due to a lower demand for electricity and a resulting increase in mains voltage. Consequently, in order to protect the packings and tubes, it is advisable to reduce the temperatures of the furnaces by 100°C for overnight stand-by.

The platinum gauze supplied with the instrument tends to break up after some time and this can result in wire fragments being drawn back along the combustion tube by the ladle; some fragments may arrive at the entrance port and eventually reach valves A and B, and consequently affect the performance of these valves. This problem can be overcome by using platinum gauze made from larger diameter wire (0·5 mm).

The need for replacement of the reduction tube packings is usually indicated by an increase in the nitrogen results. Even where three to four inches of unused copper is still present in the reduction tube, a breakthrough of oxygen can occur, which after this stage will be measured as nitrogen. Therefore, the reduction tube packing should be replaced before all copper has been converted to copper oxide and the remaining pure copper can be used again in the new packing.

In order to facilitate the removal of the oxidized copper from the tube, the authors increase the temperature of the reduction furnace to about 800°C prior to removal of the tube; after cooling, the first few inches of copper oxide are removed by means of a metal tube with a serrated end and the remainder can usually be pushed out as one fused rod.

Care should be taken when packings are replaced in order to avoid too densely or too loosely packed tubes (p. 49). A small flask vibrator assists in packing the chemicals to the correct density.

2. *Scrubbers and traps*

The frequency of replacing the packings of the scrubbers will depend upon the quality and the quantity of the gases oxygen and helium employed.

The impurity level of research grade oxygen is very low and as a consequence the packing of the scrubber will last several months. When a lower grade oxygen is used, it is advisable to remove the organic impurities by employing a pre-purifier, which converts these impurities to carbon dioxide and water; these products however are subsequently absorbed by the scrubber, thus reducing the life of its packings.

For economy and availability reasons, mineral grade helium is usually employed, which has a higher level of impurities than research grade oxygen, so scrubber replacement has to be carried out more frequently and in general it corresponds to the time the helium cylinder needs replacing.

The packings of the traps need generally replacing after 150 to 200 analyses.

Since the power to the detectors has to be switched off when replacing scrubbers and traps, it is recommended to have ready packed and purged tubes at hand, if the instrument needs to be restored for operation as quickly as possible.

B. Notes on replacement of combustion and reduction tubes

1. *Restoring deformed O-rings*

As the O-rings are deformed during use, it has been recommended[(7)] to employ new rings whenever tubes are replaced; however, the shape of the deformed O-rings can often be restored by placing them for a few minutes on top of the furnace assembly of the Model 240.

2. *Greasing of O-rings*

It has been recommended[13] to grease the O-rings lightly with a high vacuum grease, before placing them onto the tubes. However, rolling the O-rings in the palms of the hands is generally sufficient to provide a good seal and it avoids the possibility of grease entering the system.

3. *Cleaning of metal connection tube*

Occasionally the inside of the metal connection tube, situated between the combustion and reduction tubes, becomes contaminated and this can be cleaned with dilute acid, followed by thorough washing and drying.

4. *Contamination of valve D*

Sometimes the plunger of valve D may need replacing, since small particles of copper and quartz wool may enter the valve and affect its sealing. The chances of this happening can be reduced by giving particular attention to packing the reduction tube correctly.

5. *Conditioning of packing materials*

It can take a long time before the contaminants of the packing materials are removed, therefore stabilization should be carried out overnight or during the weekend.

A rapid method of conditioning the packings and replacing the combustion and reduction tubes is as follows:

(a) leave furnaces at normal operating temperatures,
(b) manually turn the programmer to position "9",
(c) remove the entrance plug to relieve the pressure in the tubes and put the ladle in a safe place,
(d) unscrew the four O-ring fittings and remove them from the ends of the tubes using asbestos gloves or tongs if necessary,
(e) withdraw the tubes from the furnaces and place them on a suitable rack for cooling,
(f) clean the ends of freshly packed tubes with chamois leather and also the tapered section of the metal fittings. Introduce the tubes into the furnace and finger tighten the O-ring fittings at the right of each tube; leave the fittings at the left disconnected,
(g) replace the entrance plug after cleaning and inspecting O-ring for damage,
(h) close the vent pipe of valve E by means of a plug or cap,

(i) manually turn the programmer through "0" to position "1" and leave for 5 minutes (helium will back-purge the reduction tube and forward-purge the combustion tube),
(j) manually turn the programmer to position "2" and leave for two minutes (helium will continue to back-purge the reduction tube and oxygen will burn off the impurities of the combustion tube packings),
(k) manually turn the programmer to position "5" and leave for two minutes (helium will remove excess oxygen from the combustion tube),
(l) manually turn the programmer to position "9" and fit the metal connector tube to the left endings of the tubes and clamp the connector in position,
(m) OPEN VENT PIPE OF VALVE E, by removing the plug or cap,
(n) manually turn the programmer to position "O" and proceed by introducing an unweighed sample for conditioning.

C. Recorder

The overall performance of the Model 240 can be affected by the condition of the recorder, therefore attention should be given to those areas which can improve the accuracy of the read-out. It is recommended to have the recorder regularly serviced either through the manufacturer or through on-site facilities.

Certain checks on the recorder, which the user can perform, will be described here.

Early Model 240 instruments used a Leeds and Northrup Speedomax W recorder, which requires regular cleaning of the pen carriage, since build-up of ink, dirt and paper dust can produce a large dead-band error. In order to effect a smooth action of the pen, a very light film of grease can be applied by running the fingers along the slide rod.

Attention should also be given to the tension of the cable, which drives the pen, and a little oil should be applied to the cable bearings and the worm gear on the servo-motor shaft. In addition adjustment should be made to the amplifier gain so that a very slight movement of the pen can be felt without recording a thick line. The setting of the damping control should give a 50% deflection without any overshoot or sluggishness and the zero setting should be checked regularly.

Later Model 240 instruments employ the Perkin–Elmer Model 056 recorder, which requires less maintenance, but attention should be given to the zero and gain settings and to the smooth action of the pen drive wire by applying oil to the bearings.

Whichever recorder is used, the pen should always be regularly cleaned and the build-up of solid material inside the pen or tubes can be reduced by withdrawing the ink back into the reservoir when the recorder is not in use.

The recorder span is usually checked during servicing or when problems are encountered in obtaining accurate results. A quick check on the recorder span can be made by means of the nitrogen or hydrogen attenuation switch of the Model 240, whereby, after checking the recorder zero setting, a 100% deflection is applied by the coarse and fine zero adjustments of the nitrogen or hydrogen bridge, while the switch is positioned on $\times 1$ attenuation. The recorder span can be regarded as correct if a deflection of exactly 25% is obtained when the nitrogen bridge output is set to $\times 4$ attenuation. In this method it is assumed that the output of the Model 240 attenuation switch is correct.

The recorder span should ideally be checked by means of a 1 mV standard input source, which will also prove useful for checking of the chart markings.

D. Modifications to the position of the programmer wheel

The programmer indicating wheel of the Model 240 is normally situated at the end of the motor driven camshaft under the electronics chassis; consequently, the cover of the instrument and the electronics chassis obstruct observation and manual turning of this wheel. Therefore, the re-

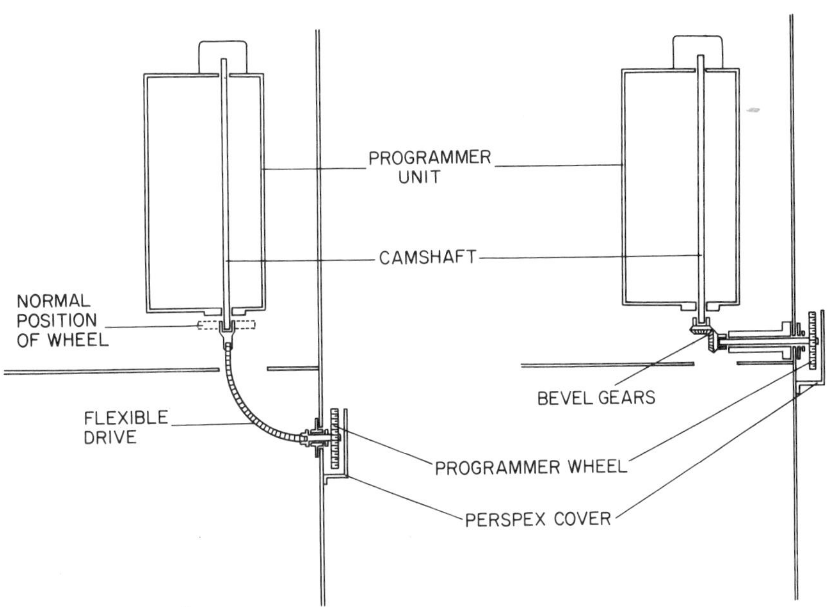

FIG. 38. Two methods of repositioning the programmer wheel.

positioning of the programmer wheel to the outside of the instrument can be advantageous during maintenance (p. 50) or when the combustion time requires extension (p. 43).

Two methods of repositioning the programmer wheel, which are known to have been used, are schematically shown in Fig. 38, and in both methods the wheel is located on the right hand side of the Model 240.

The method of using a flexible drive has the advantage that the drive mechanism unit is commercially available, but it requires the programmer to be moved to the right by about 2 cm. The use of bevel gears has been applied by D. S. Potter[30] and although this method requires more engineering, it has the distinct advantage that the programmer can remain in its original position.

Gratitude is expressed to Perkin–Elmer, Inc., for the use of some of their literature and figures, and to Roche Products, Ltd., for some photographic work.

IX. REFERENCES

1. W. Simon, P. F. Sommer and G. H. Lyssy, *Microchem. J.* **6**, (1962), 239.
2. R. D. Condon, *Microchem. J.* **10**, (1966), 408.
3. R. F. Culmo, *Mikrochim. Acta,* (1968), 811.
4. R. F. Culmo, *Microchem. J.* **17**, (1972), 499.
5. R. F. Culmo, J. Sibrava, L. VanBey and A. P. Gray, Pittsburgh Conference on Analytical Chemistry and Applied Spectroscopy, Cleveland, Ohio, U.S.A., March 1975.
6. A. Steyermark, "Quantitative Organic Microanalysis", 2nd Edn, Academic Press, New York and London (1961).
7. Perkin–Elmer Manual (1972).
8. J. Grant, "Quantitative Organic Microanalysis, based on the methods of Fritz Pregl", 4th Edn, Churchill, London 1945.
9. G. Ingram, "Organic Elemental Analysis", Chapman & Hall, London, 1962.
10. K. P. Kunz, *Microchem. J.* **13**, (1968), 463.
11. P. C. Nicolson and F. Puzio, "Proceedings of the Sixth International Symposium on Microtechniques, Graz, Austria", Verlag der Wiener Medizinischen Akademie, 1971, p. 231.
12. A. M. G. Macdonald and G. G. Turton, *Microchem. J.* **13**, (1968), 1.
13. G. Gustin and M. L. Tefft, *Microchem. J.* **10** (1966), 236.
14. H. J. Francis, Jr., *Microchem. J.* **7** (1963), 462.
15. C. S. Yeh, *Microchem. J.* **7** (1963), 303.
16. R. F. Culmo, *Mikrochim. Acta,* (1969), 175.
17. R. F. Culmo and R. Fyans, *Mikrochim. Acta,* (1968), 816.
18. G. E. Secor and L. M. White, *Anal. Chem.* **38** (1966), 945.
19. R. Belcher, J. E. Fildes and A. J. Nutten, *Anal. Chim. Acta,* **13** (1955), 431.
20. H. Swift, *Microchem. J.* **11** (1966), 193.
21. R. F. Culmo, "Elemental Analysis Application Study I", Perkin–Elmer, January 1974.

22. Y. A. Gawargious and A. M. G. Macdonald, *Anal. Chim. Acta,* **27** (1962), 300.
23. Y. A. Gawargious and A. M. G. Macdonald, *Anal. Chim. Acta,* **27** (1962), 119.
24. J. G. Stockner and F. A. J. Armstrong, *J. Fish. Res. Board of Canada,* **28** (1971), 215.
25. D. W. Menzel and J. H. Ryther, *Deep Sea Research,* **15** (1968), 327.
26. L. A. Hobson and D. W. Menzel, *Limnology and Oceanography,* **14** (1969), 159.
27. R. K. Patterson, *Anal. Chem.* **45** (1973), 605.
28. A. J. Smith, G. Myers, Jr. and W. C. Shaner, Jr., *Mikrochim. Acta,* (1972), 217.
29. H. Wagner, *Perkin–Elmer Instrument News,* **20** (1970), No. 4.
30. Society for Analytical Chemistry, London, Microchemical Methods Group, Elemental Analyzer User Forum, *Newsletter* **1** (1974).

2. THE PERKIN–ELMER MODEL 240 ELEMENTAL ANALYZER: OXYGEN OR SULPHUR

M. R. Cottrell and F. H. Cottrell (née: van der Voort)
Roche Products Ltd, Welwyn Garden City
and
Hatfield Polytechnic, Hatfield, Hertfordshire

I. INTRODUCTION

The Perkin–Elmer Model 240 elemental analyzer, which was originally designed to determine carbon, hydrogen and nitrogen in organic compounds, has proved itself capable by its versatility of analysing these elements in a very wide range of compounds, and in later developments this basic instrument was adapted to determine in addition oxygen(1) and sulphur.(2)

The conversion of the Model 240 to the determination of oxygen was introduced in 1967 and it involves replacement of the combustion and reduction tubes by pyrolysis and oxidation tubes with the addition of a U-shaped absorption tube fitted between them. The conditions for the oxygen analysis were based on the modified Schütze method of Oita and Conway,[3] and while the accuracy of the results by the Model 240 equals the classical procedure, the speed and simplicity of the automatic method is an important improvement.

A similar conversion of the CHN-analyzer to determine sulphur was introduced in 1972; this also involves replacement of the oxidation and reduction tubes and the addition of a U-shaped absorber tube, while a further replacement of the water-absorption tube by a heated sulphur dioxide absorber is required. After empirically changing the conditions of packings and temperature, certain reproducible conditions were found to convert the sulphur completely to sulphur dioxide, which forms the basis of the sulphur determination. In addition to sulphur, carbon and nitrogen can often be determined simultaneously under these conditions.

II. DETERMINATION OF OXYGEN

A. Modification to the Model 240

The Perkin–Elmer CHN-analyzer can be adapted to the direct micro-determination of oxygen in organic compounds by replacing the combustion and reduction tubes respectively by pyrolysis and oxidation tubes.[1] In addition an acid-gas scrubber tube is connected between the outlet of the pyrolysis tube and the inlet of the oxidation tube and for obvious reasons the oxygen inlet valve (valve B) of the Model 240 is disconnected.

The pyrolysis and oxidation tube layout in the Model 240 is shown diagrammatically in Fig. 1.

Perkin–Elmer supply a conversion kit, which provides a quick and easy conversion of the components from the CHN-mode of the Model 240 to the oxygen mode.

B. Chemistry of pyrolysis, oxidation and absorption

The history of the development of the direct determination of oxygen has been summarized by Belcher *et al.*,[4] who studied the various modifications described in the literature.

The direct determination of oxygen by means of the modified Model 240 elemental analyzer was mainly based on the modified Schütze method of Oita and Conway[3] and accordingly the chemistry involved is similar.

The organic compound is pyrolyzed in a helium atmosphere at a temperature of 975°C in the presence of platinized carbon, and under these con-

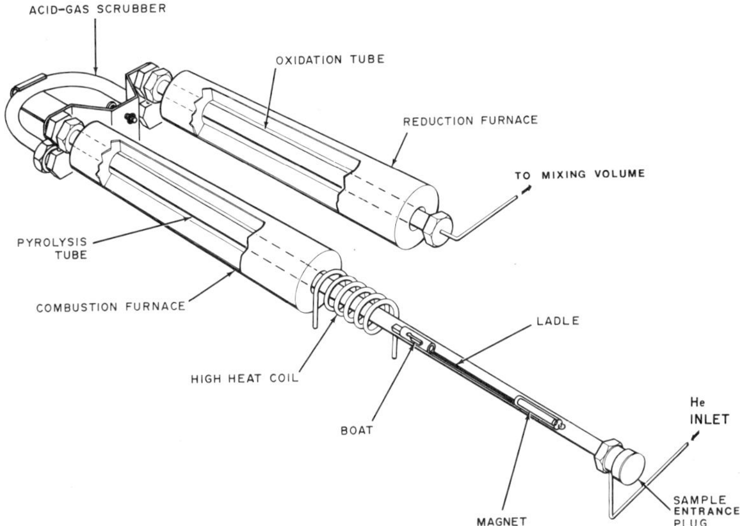

FIG. 1. Lay-out of pyrolysis, oxidation and scrubber tubes in the oxygen mode.

ditions the oxygen is converted to carbon monoxide. Interfering pyrolysis products are either removed by hot copper (900°C) or absorbed in a U-shaped scrubber usually containing lithium or sodium hydroxide.

The carbon monoxide is subsequently passed through copper oxide at 670°C in the oxidation tube, where it is converted to carbon dioxide, which is finally absorbed by lithium or sodium hydroxide in the CO_2-trap of the carbon detector bridge.

C. Tubes, packings and fillings

Most of the reagents required in the oxygen analysis are also used in the Model 240 CHN-analysis, with the exception of platinized carbon, which plays an important role in the oxygen determination.

The manufacturers supply recommended chemicals with the conversion kit. Other reagents may give equally satisfactory performance, but care should be taken to avoid damage to the instrument.

1. *Pyrolysis tube*

The pyrolysis tube is similar to the combustion tube used in the CHN-analysis and it is also manufactured from high quality transparent quartz. Its total length is 25 inches, of which 12 inches are inside the furnace and 8 inches at the end of the heated part are occupied by the pyrolysis tube packings. The packings are located between a small roll of platinum gauze and a roll of silver gauze at the end of the tube, the respective catalytic or absorbing capacity of these metals contribute during the pyrolysis of the sample. The temperature of the furnace for the pyrolysis is maintained at 975°C. The packing arrangement of the pyrolysis tube is shown in Fig. 2.

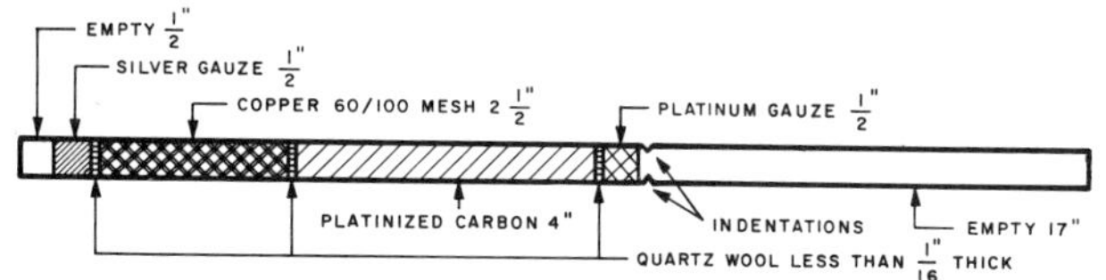

FIG. 2. Packings of the pyrolysis tube.

Platinized carbon at 975°C is regarded as an efficient reagent(3,4) in effecting the conversion of oxygen to carbon monoxide. For this purpose the manufacturers supply 50% platinized carbon, which is manufactured as described by Oliver.(5)

The heated copper is included in this packing for the removal of some interfering sulphur products as recommended by Oita and Conway.

The individual reagents are separated by small plugs of quartz wool.

2. *Acid–gas scrubber tube*

The U-shaped absorption tube is located between the pyrolysis and oxidation tubes by means of special O-ring connectors and the tube is held in position by a retaining bracket. The tube is filled with lithium or sodium hydroxide for the removal of acid pyrolysis products and quartz wool plugs are placed at each end.

The dimensions of the U-tube and its packings are shown in Fig. 3.

3. *Oxidation tube*

The oxidation tube, which is maintained at 670°C is similar to the reduc-

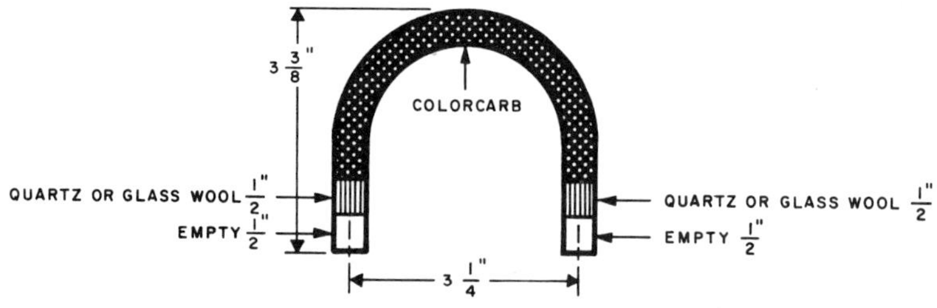

FIG. 3. U-shaped scrubber tube and packings.

tion tube of the CHN-analyzer and its packings are located by means of small rolls of silver gauze arranged as shown in Fig. 4.

Heated copper oxide is the active ingredient of the tube to convert carbon monoxide to carbon dioxide. It is situated between two packings of heated copper, while small plugs of quartz wool are used for separation of the individual packings.

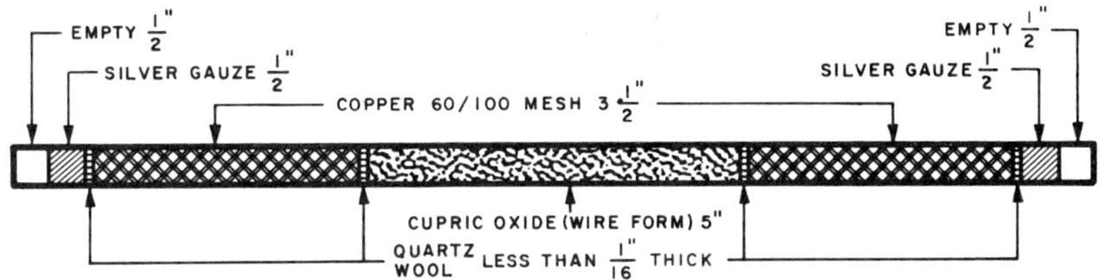

FIG. 4. Packings of the oxidation tube.

4. *Carbon dioxide absorption tube*

The carbon dioxide produced during the oxidation stage eventually reaches the carbon detector bridge, where it is absorbed by lithium or sodium hydroxide. The fillings of this absorption tube are identical to those used for the absorption of carbon dioxide in the CHN-analysis and are shown in Fig. 5.

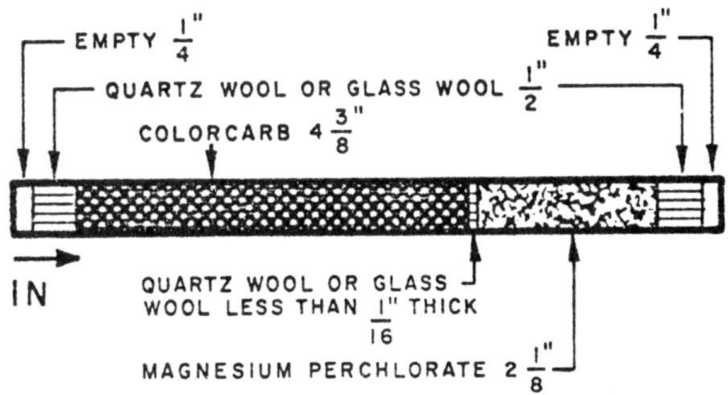

FIG. 5. Packings of the carbon dioxide absorption tube.

D. Conversion and procedure

Detailed instructions[6] on the conversion and operation of the Model 240 for oxygen analysis are provided by Perkin–Elmer with the conversion kit, therefore the following discussion will be limited to basic information only. Sample handling techniques are the same as those described in the CHN-chapter.

1. *Conversion and preparation of the Model 240*

In order to safeguard the detector filaments, it is important to switch the instrument off or turn the programmer to position "9", before converting the instrument.

The combustion and reduction tubes are removed from the Model 240 and respectively replaced by the pyrolysis and oxidation tubes, which should be packed as described before. The U-shaped scrubber tube, which is packed with lithium or sodium hydroxide, is positioned by means of O-ring sealed connectors between the pyrolysis and oxidation tubes and it is secured by a retaining bracket.

The conversion is completed by disconnecting a lead from the solenoid coil of valve B, which stops the operation of this valve.

2. *Conditioning of the packings*

A correct gaseous equilibrium within the Model 240 is essential for the achievement of accurate results and it is even more critical for the analysis of oxygen than it is for the CHN-analysis, since the equilibrium within the platinized carbon can be influenced by a small change of temperature as well as by the flow of helium constantly purging through it.

The instrument is conditioned, therefore, by pyrolyzing several unweighed samples, while employing the same procedure as for a normal analysis. This will establish a correct equilibrium between the gases flowing through the instrument and those absorbed onto the chemicals and components.

These conditioning runs can be utilized to adjust the setting of the helium inlet pressure to a level which allows the mixing volume to reach a pressure of 1500 mmHg in $3\frac{1}{2}$ minutes ($\pm\frac{1}{2}$ minute).

3. *Blank determinations*

Reproducible blank values depend on the correct conditioning of the instrument and after each blank determination another conditioning run by means of an unweighed sample, will have to be carried out to re-establish the correct equilibrium.

It has been suggested that blank values can be reduced by switching to the "hold" position 30 seconds after the start button has been pushed, which allows helium to purge the air, introduced during the ladle insertion, from the pyrolysis tube. If this method is used for the determination of the blank value, the same procedure should be followed consistently during all following analysis.

4. *Calibration of the instrument*

A standard compound is used to establish the relationship between the output signal of the carbon detector bridge, produced by the carbon dioxide from the oxidation stage, and the weight of oxygen in the original compound. The resulting constant relates the number of microvolts detected to each microgram of oxygen and a typical value for oxygen is 16 μV/μg.

Microanalytical grade benzoic acid is the standard recommended by Perkin–Elmer for use in the calibration procedure, but the final accuracy of the results may be improved if a standard is used which matches the composition of the sample to be analyzed, and a wide range of pure organic compounds can be used for this purpose.

5. *Procedure of operation*

The procedure employed for the determination of oxygen in the Model 240 is very similar to that of the CHN-analysis, which is described in the previous chapter, and varies only in the time at which the sample is introduced into the furnace and the length of time during which helium purges the pyrolysis products through the packing materials.

About 1–3 mg of sample is accurately weighed into a clean platinum boat and entered into the end of the quartz ladle, which is then inserted into the cool section of the pyrolysis tube. The combust button is pressed $2\frac{1}{4}$ minutes after starting the analysis, which is indicated by the inject light and the ladle is subsequently introduced into the pyrolysis furnace at position $4\frac{1}{4}$ into the programmed analysis cycle, at which time valve A (the helium valve) will be open allowing the pyrolysis products to be purged through the packings and into the mixing volume. The helium inlet pressure has been previously set to allow the pressure inside the mixing volume to reach 1500 mmHg within $3\frac{1}{2}$ minutes, upon which valve D closes. The analysis sequence is automatic from this point onwards and it will result in a signal at the carbon detector bridge, which may need suppressing if the signal is too large.

The final signal is then related to the sample weight and the calibration constant in order to calculate the percentage of oxygen in the sample. The calculations involved are identical to those used for the carbon analysis, except that oxygen is substituted for carbon.

It should be noted, however, that signals recorded for hydrogen and nitrogen may not be meaningful, since under pyrolysis conditions various interelement compounds,[(4)] such as HCN may be formed.

E. Additional information

1. *Speed of conversion from CHN to oxygen*

The components conversion of the Model 240 from CHN to oxygen analysis can be achieved in 10 to 15 minutes, if the various tubes are packed ready for installation. Attaining equilibrium within the instrument, however, can take a long time, particularly when new packings are employed.

It is therefore recommended to accumulate sufficient samples for several days of operation in the oxygen mode, or if possible to assign one instrument solely to oxygen determinations.

In order to avoid unnecessary changes to the equilibrium within the pyrolysis tube when conversions are performed regularly, it is recommended to seal the packed tube in a helium atmosphere during storage.

2. *Applications*

Culmo[1] reported the oxygen results obtained for various organic compounds, containing carbon, hydrogen, nitrogen, chlorine, bromine, iodine, and sulphur, which easily satisfied the 0·3% accuracy required.

Smith *et al.*[7] determined oxygen in typical petroleum compounds by means of the Model 240 as well as a modified Schütze method,[8] and the results obtained by the Model 240 were in good agreement with the classical procedure.

It appears from discussions with Model 240 users that sulphur containing samples sometimes provide problems even though copper is present in the pyrolysis tube. Refractory compounds such as organometallics or boron-containing samples, usually prove difficult to decompose and compounds containing fluorine provide problems due to attack on the quartz.

The problems discussed above have also been experienced in the classical procedures[9] to which the reader is referred, since as yet little has been published on the oxygen determination by the Model 240 analyzer.

III. DETERMINATION OF SULPHUR

A. Modification to the Model 240

The Perkin–Elmer elemental analyzer can also be adapted to perform direct microdeterminations of sulphur in organic compounds.[2] The modification to the CHN-analyzer consists of replacing the standard combustion and reduction tubes by differently designed tubes containing specific reagents for the determination of sulphur. In addition a halogen-gas scrubber tube is connected between the outlet of the combustion tube and the inlet of the reduction tube.

The diagrammatic layout of the combustion and reduction tubes for the sulphur analysis is shown in Fig. 6.

Furthermore, the water absorption tube, which is situated on the outside

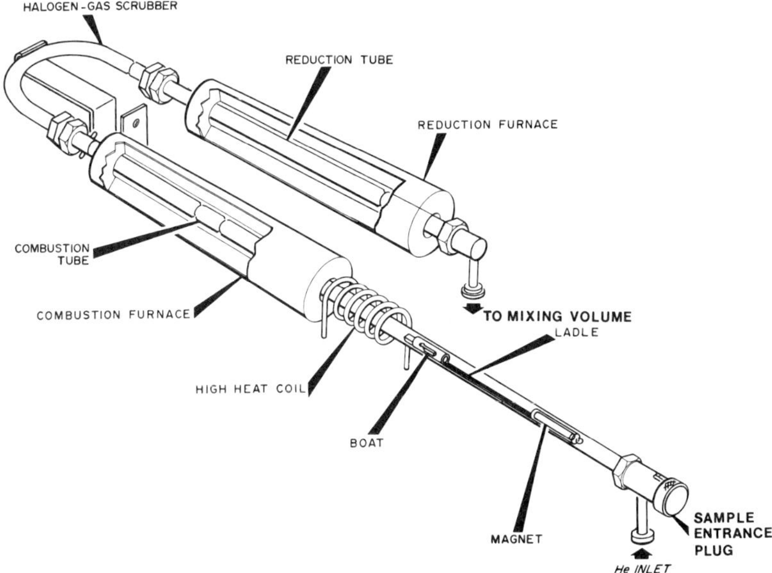

FIG. 6. Lay-out of combustion, reduction and scrubber tubes in the sulphur mode.

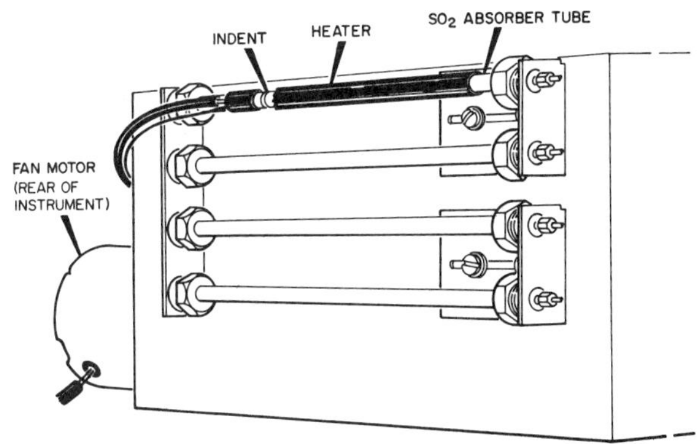

FIG. 7. Sulphur dioxide absorption tube and heater.

of the oven unit, is replaced by a sulphur dioxide absorption tube surrounded by a small heater as is shown in Fig. 7.

By employing the sulphur conversion kit available from Perkin–Elmer, the components of the CHN-analyzer can be replaced fairly quickly.

B. Chemistry of combustion, reduction and absorption

The direct determination of sulphur by the Model 240 utilizes established reagents, but unconventional packing procedures, which apparently achieves a quantitative conversion of the sulphur from the organic compound to the single species sulphur dioxide.

The method was arrived at after much experimentation by Culmo,[2] who applied his experience with tungstic oxide as an oxidation catalyst for sulphur[10] to the Model 240 as a basis. By empirically changing the conditions of packings and temperature in the combustion and reduction tubes, Culmo found certain critical conditions which satisfied a complete conversion to sulphur dioxide, and in addition the determination of carbon and nitrogen could be obtained.

The areas studied in this empirical approach were the formation of sulphur trioxide, water, and the effect on the gaseous mixture by the copper in the reduction tube. The resulting optimal conditions involved tightly packing the tungstic oxide in the combustion tube and "channelled" copper in the reduction tube, which caused the gases to remain longer in contact with the tungstic oxide in the oxygen atmosphere of the combustion zone (975°C) than with the copper in the reduction zone (820°C).

Water formed during the combustion is absorbed by either magnesium perchlorate or calcium chloride, situated inside the combustion tube but outside the furnace, and residual halogen byproducts are removed by 8-hydroxyquinoline in the U-shaped scrubber tube.

The final gaseous products of the analysis are passed to the detectors, where sulphur dioxide is absorbed by silver oxide at 210°C, and carbon dioxide by lithium or sodium hydroxide, while nitrogen is measured against helium.

C. Tubes, packings and fillings

Because this sulphur method is based on many empirical factors, it is advisable to employ only the reagents recommended and supplied by Perkin–Elmer in order to obtain satisfactory performance of the instrument.

It is important to adhere strictly to the detailed instructions[11] provided by Perkin–Elmer with the sulphur analysis kit, since successful analyses are dependent on the special packing procedures employed.

Magnesium perchlorate is a very strong oxidant; as it is employed in the heated combustion tube directly outside the furnace, it presents an explosive hazard. Every care should therefore be taken that this reagent stays outside the heated furnace. In this respect calcium chloride is a safer reagent, but it is less efficient.

1. *Combustion tube*

The combustion tube is manufactured from high quality transparent quartz and it has three sets of indentations for positioning of the packings. The packing arrangement of the tube is shown in Fig. 8.

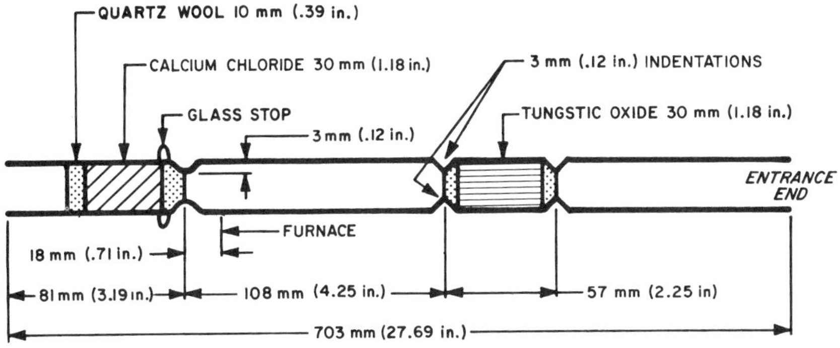

FIG. 8. Packings of the combustion tube.

The only packing inside the furnace (975°C) is tungstic oxide, which is tightly packed between quartz wool plugs and positioned between two sets of indentations in the tube. The following section up to the next set of indentations remains empty, and calcium chloride or magnesium perchlorate between quartz wool plugs is the other packing of the tube, which is prevented from entering the heated area by glass stops for safety reasons.

2. *Halogen-gas scrubber tube*

The halogen-gas scrubber tube is located between the outlet of the combustion tube and the inlet of the reduction tube by means of special O-ring connectors and it is held in position by a retaining bracket. The tube is filled with 8-hydroxyquinoline between plugs of quartz wool as is shown in Fig. 9.

3. *Reduction tube*

The reduction tube is maintained at 820°C, and is manufactured from quartz with one set of indentations for positioning of the packing. The reduction tube and packings are diagrammatically shown in Fig. 10.

Wire-form copper is loosely packed between quartz wool plugs, and extends to approximately seven inches from the set of indentations near the centre of the tube. As a result of this loose packing an empty channel of 1–3 mm is created over the total length of the copper. The reduction packing removes residual oxygen, reduces nitrogen oxides to nitrogen, and ensures that

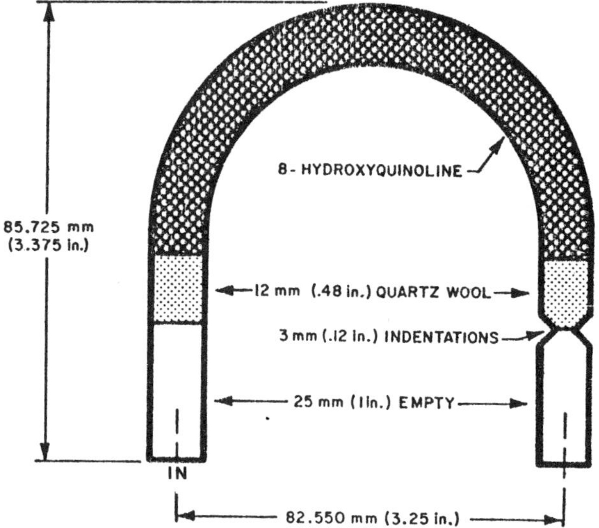

FIG. 9. U-shaped tube and packings.

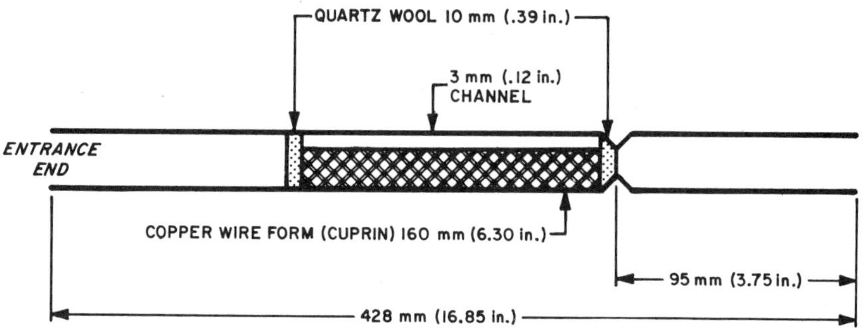

FIG. 10. Packings of the reduction tube.

finally sulphur dioxide, carbon dioxide and nitrogen are transferred to the detection stage.

4. *Sulphur dioxide absorption tube*

The sulphur dioxide absorption tube and its heater (210°C) replace the water absorption tube on the outside of the oven unit; the tube is shown diagrammatically in Fig. 11.

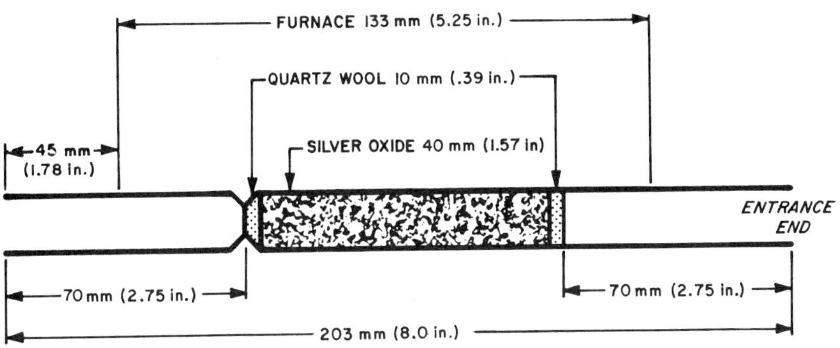

FIG. 11. Packings of the sulphur dioxide absorption tube.

The filling in the tube is silver oxide, which is packed as normal between plugs of quartz wool.

The carbon dioxide absorption tube is the same as that used for the CHN-analysis.

D. Conversion and procedure

Perkin–Elmer provide instructions[11] with the sulphur analysis kit on the conversion and operation of the Model 240, which should be studied for detailed information on the methods employed. This discussion will be limited to basic information only. Sample handling techniques are the same as those described in the CHN chapter.

1. *Conversion and preparation of the Model 240*

The conversion involves an interruption of the gas flow through the detectors; therefore, in order to safeguard the detector filaments, the instrument should be switched from the "detect" to the "on" position before the conversion.

After removing the CHN combustion and reduction tubes, the ready-packed sulphur combustion tube is introduced from the left until the glass stops of the tube reach the furnace, which avoids accidental introduction of the magnesium perchlorate into the hot furnace. After tightening the O-ring connection at the sample inlet end of the tube and setting the programmer to position "2", the flow of oxygen at the outlet of the tube is adjusted by the oxygen inlet pressure to approximately 80 ml/min on a flow meter.

The reduction and U-shaped scrubber tubes are positioned and sealed by means of O-ring connectors.

The sulphur dioxide absorption tube inside its heater unit replaces the water absorption tube of the CHN-analyzer and after switching the analyzer off, the power and the thermocouple leads of the heater unit are connected to the Model 240.

For the sulphur analysis, Perkin–Elmer recommend a higher helium inlet pressure. However, the resulting increase in the flow rate of the helium through the detectors may cause tailing of the sulphur dioxide signal, which can be reduced by means of a restrictor tube fixed onto the outlet of the instrument; in addition the increased helium pressure will cause the programmer to run through the inject stage, which removes the need for pressing the combust button.

2. *Conditioning of the system and blank determinations*

In order to obtain a correct gaseous equilibrium within the Model 240, it is necessary to condition the packings and components of the instrument.

The method of conditioning suggested by Perkin–Elmer[11] starts with purging helium through the system, which removes the atmospheric nitrogen and trace moisture. This is followed by blank determinations until the carbon and sulphur blanks are less than 300 μV. Finally an unweighed sample is taken through the normal analysis procedure for completion of the conditioning of the system.

In the authors' opinion a blank value should again be determined after the final conditioning run, which would be more representative of analysis conditions. Some additional information on conditioning is given below.

The conditioning runs can be utilized to adjust the setting of the helium inlet pressure to the recommended level, which allows the mixing volume to reach the fixed pressure limit of 1500 mmHg in 30 to 60 seconds.

3. *Calibration of the instrument*

The relationship between the output signals of the detector bridges and the weight of the elements to be analysed is established by means of a standard compound. The resulting constants relate the number of microvolts detected to each microgram of element and a typical value for sulphur is 18 μV/μg.

The standard recommended by Perkin–Elmer is phenylthiourea, but the final accuracy of the results may be improved if the standard matches approximately the composition of the sample to be analyzed.

4. *Procedure of operation*

The procedure employed for the determination of sulphur in the Perkin–Elmer elemental analyzer is similar to that for the CHN-analysis, which is described in the previous chapter; only the time at which the sample is introduced into the furnace and the time during which helium purges the combustion products through the packing materials are varied.

If a sample which is easy to combust is to be analysed, 1 to 2 mg of sample is accurately weighed in a clean platinum boat and placed inside the quartz ladle, which is subsequently inserted into the cool section of the combustion tube. The ladle is introduced into the hot furnace at position 4 into the programmed analysis cycle, which is just before valve A (the helium valve) opens.

When analyzing refractory compounds, tungstic oxide should be added to the 1 to 2 mg sample in the platinum boat before it is placed into the ladle and subsequently into the cool section of the combustion tube. The ladle is introduced into the hot furnace when the inject light switches on at position $2\frac{1}{4}$ into the programmed cycle, in order to provide extra time for combustion.

It is essential not to exceed this 1–2 mg range for sample weight, since the possibility of incompletely burned products coming in contact with the magnesium perchlorate could be dangerous (p. 68).

The subsequent analysis cycle will be automatic and it will result in signals at the detectors, which may need attenuation if the signals are too large. The final signals are then related to the sample weight and the calibration constants in order to provide the percentages of the elements present in the original sample. The final calculations are identical to those used in the CHN-analysis, except that sulphur is substituted for hydrogen.

The conditions used in this method were deliberately aimed at optimizing the analysis for sulphur, but although most analyses will provide excellent results for all three elements nitrogen, carbon and sulphur, the results for nitrogen and carbon occasionally fall outside the accepted 0·3% accuracy.

E. Additional information

1. *Speed of conversion from CHN to sulphur*

Converting the components of the Model 240 from CHN to sulphur analysis can be achieved in 20 to 30 minutes, but stabilizing the detectors and oven temperatures will take significantly longer. It is recommended therefore, that the conversion is carried out at the end of the day, which allows the instrument to stabilize overnight.

The time needed for conditioning the instrument will depend partly on the purity of the packings, which can be considerably improved if the reagents employed are preconditioned. Silver oxide can be conditioned by heating a bulk quantity at 200°C for 48 hours, tungstic oxide by heating it at 900°C for one hour and wire-form copper by heating it under flowing helium for 2 hours at 400°C. All reagents are allowed to cool at room temperature before they are stored in a desiccator.

It is further advisable to preheat (400–500°C) the ready-packed reduction tube in flowing helium, before installing it into the Model 240. In addition, the impurities from the quartz wool can be removed by heating the wool for a short time over a flame.

Another method of conditioning the packings has been described in the CHN-analysis (p. 53); this involves back-purging the heated reduction tube with helium and forward-purging the heated oxidation tube with sequentially oxygen and helium, while the tubes are in position in the Model 240.

The above described preconditioning methods will prevent, for example, premature exhaustion of the small amount of copper in the sulphur reduction tube by the many blanks and conditioning runs, which would otherwise be required for stabilizing the system.

2. *Applications*

Culmo[2] determined sulphur in a wide range of organic and inorganic compounds by the Model 240 and the reported results were all within the required 0·3% accuracy.

Further results for sulphur as well as carbon and nitrogen for a variety of compounds were reported by Culmo[12] and an excellent agreement with the theoretical values was shown.

Gratitude is expressed to Perkin–Elmer, Inc., for the use of some of their literature and figures, and to Roche Products, Ltd., for some photographic work.

IV. REFERENCES

1. R. F. Culmo, *Mikrochim. Acta,* **4** (1968), 811.
2. R. F. Culmo, *Microchem. J.* **17** (1972), 499.
3. K. J. Oita and H. S. Conway, *Anal. Chem.* **26** (1954), 600.
4. R. Belcher, D. H. Davies and T. S. West, *Talanta,* **12** (1965), 43.
5. F. H. Oliver, *Analyst,* **80** (1955) 593.
6. Perkin–Elmer Publication 990–9630, May 1968.
7. A. J. Smith, G. Myers Jr. and W. C. Shaner Jr., *Mikrochim. Acta* **2** (1976), 217.
8. M. Dundy and E. Stehr, *Anal. Chem.* **23** (1951), 1408.
9. A. Steyermark, "Quantitative Organic Microanalysis", 2nd Edn, Academic Press, New York and London, 1961.
10. R. C. Rittner and R. F. Culmo, *Microchem. J.* **11** (1966), 269.
11. Perkin–Elmer Publication 990–9883, Feb. 1973.
12. Perkin–Elmer *Instrument News,* **22** (1972), No. 1E.

3. THE CARLO ERBA ANALYZER: CARBON, HYDROGEN, NITROGEN AND OXYGEN

C. J. HOWARTH

I.C.I. Ltd., Pharmaceuticals Division, Alderley Park, Macclesfield, Cheshire

I. INTRODUCTION

The analysis of compounds for carbon, hydrogen and nitrogen by manual methods based on the work of Pregl requires two separate stages: one for the determination of carbon and hydrogen (usually by combustion in a stream of oxygen) and the other for the determination of nitrogen.

The carbon, hydrogen and nitrogen elemental analyzers introduced in the early nineteen sixties determined the three elements with one analysis. These instruments gained favour because of their speed, precision, and the reduction of the amount of sample required for a CHN analysis. In addition there is a reduction in the specialist training required compared with the Pregl type methods. This is a big step forward but whilst the analysis is automatic the instruments are only semi-automatic, for each sample has to be introduced separately by an operator.

The introduction of the Carlo Erba CHNO analyzer represented a change to a fully automatic commercial elemental analyzer. This was the first of what may be called the "second generation" instruments because the presence of an operator is unnecessary once a "run" or magazine of samples has been started.

The main consideration will be the use of the Carlo Erba as an analytical instrument rather than the chemistry behind it which has been studied by Pella and Colombo for both oxygen analyses[(1)] and for CHN analyses.[(2)]

II. THE PRINCIPLE OF THE AUTOMATION AND THE COMBUSTION

The Carlo Erba operates on a gas chromatographic system whereby nitrogen, carbon dioxide and water vapour from the dynamic combustion of organic samples are eluted with helium and detected by a catharometer. Results are recorded with a 1 mV recorder and areas under the peaks are integrated electronically and printed out.

Whereas in the semi-automatic analyzers the samples are introduced manually into the combustion furnace, the Carlo Erba enables 23 samples to be weighed and dropped in sequentially by means of a pneumatically actuated sample dispenser. There is no need to open the system to the atmosphere between each analysis; hence there is no need to wait until the air peak has passed through. This saves time, only 10–12 minutes being required for a CHN analysis and 5–7 minutes for an oxygen analysis.

There are two separate channels, one for CHN and the other for oxygen. Simple switching enables 23 oxygen analyses to be conducted immediately after 23 CHN analyses. This is made possible by means of a second sample dispenser on the oxygen channel. The CHN or oxygen channels are used as reference channels for each other. The schematic diagram for both channels is shown in Fig. 1.

1. *Combustion–CHN channel*

The weighed samples, encapsulated in tin, are introduced into the combustion zone at 1050°C to coincide with oxygen enrichment of the helium

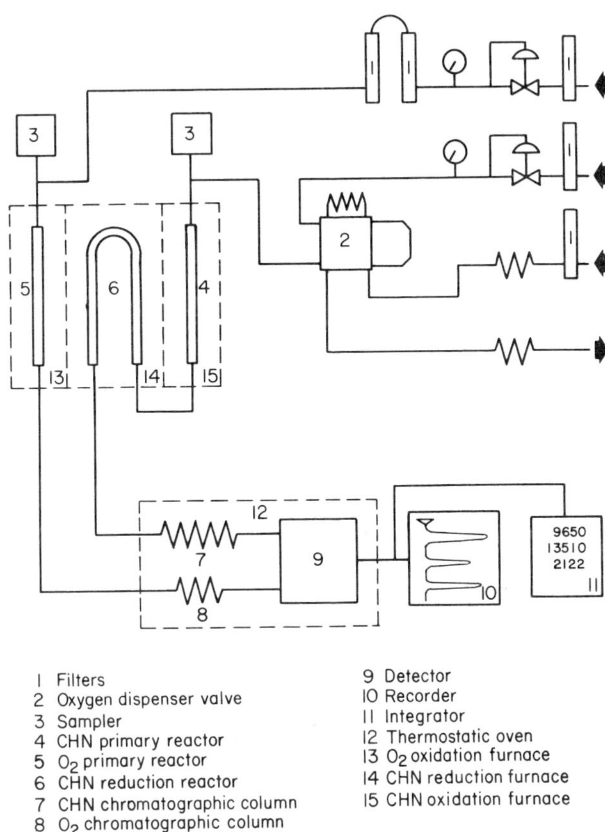

FIG. 1. Schematic diagram for CHN and O channels.

carrier gas. This facilitates the combustion which is completed by the passage of the combustion products through the combustion tube. The carbon is converted quantitatively to carbon dioxide, the hydrogen to water and the nitrogen to oxides of nitrogen; halogen and sulphur products, if present, are absorbed on silver.

A secondary reaction tube removes excess oxygen and reduces the nitrogen oxides to nitrogen before the gases enter the chromatographic column.

2. *Pyrolysis oxygen channel*

The oxygen analysis is similar except that silver capsules are used instead of tin. The weighed samples are instantaneously pyrolyzed in the helium stream and the oxygen-containing gases are quantitatively converted to carbon monoxide by contact with a special form of carbon at 1000°C.

The other main products of pyrolysis are nitrogen, traces of methane and hydrogen, but the pyrolysis of organic compounds in an inert gas is highly complex.[(3)]

After removal of acidic combustion products the carbon monoxide is separated chromatographically and measured quantitatively.

III. THE APPARATUS—DESCRIPTION OF THE VARIOUS PARTS

The complete instrument, Fig. 2, consists of three separate units namely an analytical unit, a control unit and an integrator. They will be fully described to help with the understanding of parts of the notes at the end of the chapter.

A. The analytical unit

This includes all the equipment necessary for the introduction and efficient combustion of the sample, and also for the preparation, separation and quantitative measurement of the combustion products.

1. *Combustion trains*

The CHN combustion train consists of a vertical combustion tube connected in series to a metallic U-shaped reduction tube. The tubes are interconnected by means of compression joints sealed with Viton O-rings. The oxygen combustion train consists of a single vertical quartz tube. Both combustion tubes are mounted together with the CHN reduction tube in a multizone furnace. The CHN tube is normally at 1050° ±10°C and the oxygen tube at 1000°C. This is possible because each combustion tube has its

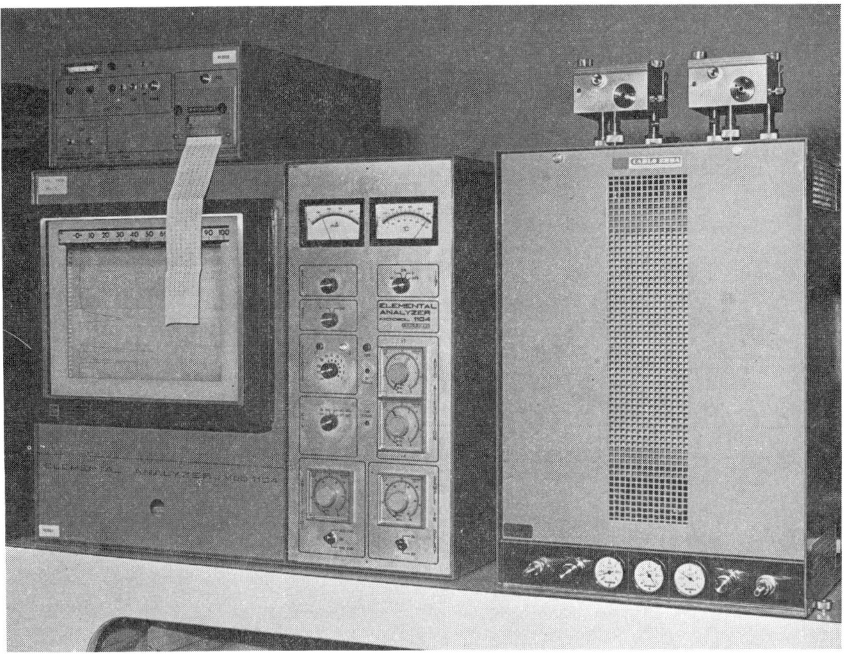

FIG. 2. Complete instrument showing the integrator, recorder control unit and analytical unit.

own heating element, temperature control and display. The residual heat from the combustion tubes is used to maintain the temperature of the reduction tube which is also displayed. This is controlled at 600–650°C by means of a variable vent on top of the multizone furnace.

The upper parts of both combustion tubes are surrounded by water-cooled coils to prevent heat transfer to the automatic samplers.

2. *Automatic sampler*

An automatic sampler, Fig. 3, is mounted above the combustion tube and consists of a removable rotating drum around whose circumference are 24 5-mm diameter holes numbered 0, 1, . . . , 23. The drum is housed in a leakproof metal body closed by a cover in which there is a transparent window directly above the combustion tube. The system is provided with a purging valve at the front.

The whole device is mounted on the instrument by means of three drilled legs; again, O-rings and locking nuts provide gas tight seals. The helium carrier gas flows through two of the legs and compressed air through the third. The drum, like a revolver magazine is rotated step by step and at

FIG. 3. Automatic sampler and removable magazine.

each displacement a container falls through a carrier leg into the combustion tube. The rotation of the drum is controlled by a piston actuated automatically by means of a compressed air servomotor. It may also be operated manually by a lever on the side of the sampler.

Automatic samplers may be fitted to both CHN and O channels so that either channel is ready for use. If only one sampler is used the carrier gas flow is provided in the second analytical channel through a metal loop with locking nuts.

3. *Chromatographic columns and the detector*

The chromatographic columns for the separation of the combustion products are enclosed together with the detector, in a thermostated oven. The oven temperature is maintained at 120°C by an electronic temperature controller having as sensing element a platinum resistance thermometer.

The thermal conductivity detector consists of a stainless steel block housing four tungsten–rhenium filaments. These four filaments are mounted in a Wheatstone bridge resistance circuit; two arms are in the CHN channel, and two arms in the oxygen channel.

4. *Pneumatic circuits*

Separate inlets for the carefully controlled flow of high purity GC grade helium carrier gas are provided for the CHN and O channels. Both circuits include the following: an inlet connector for a filter, a pressure controller with a 3 kg/cm^2 inlet pressure, an outlet pressure adjustable from 0·1 to 0·3 kg/cm^2, and a flow restrictor gauge showing the back pressure of the analytical system. The inlet filter for the CHN train is filled with $CuSO_4 \cdot 5H_2O$. In addition the CHN train contains an oxygen loop which enables a fixed quantity of oxygen to be introduced into the carrier gas stream by means of a pre-set automatic injection valve.

The auxiliary oxygen supply for the CHN circuit contains a filter for the removal of traces of carbon dioxide and water before the inlet connector, and includes a switching valve with an injection loop, an outlet connector and a needle valve to regulate the oxygen flow into the system.

In the oxygen channel there are two filters, one before the inlet to remove carbon dioxide and water, and the other between the combustion tube and the chromatographic column to remove acidic combustion products.

The analytical unit also contains the servo gas pressure controller allowing a 3 kg/cm^2 inlet pressure and an outlet pressure adjustable from 0·1 to 3 kg/cm^2. Included in this circuit are two 3-way solenoid valves controlling the automatic samplers of the CHN and O channels respectively, and the 4-way solenoid valve which controls the switching valve for the addition of oxygen to the carrier gas.

This valve consists of a Neoprene membrane held between two metallic blocks. In the rear block, which is the control section, there are eight cells connected alternately four by four. Each set of four cells is connected to separate ports on a four-way solenoid valve leaving the remaining two ports on this valve to be used for the compressed air.

B. The control unit

This contains all the timers, switches and indicators necessary for the setting and monitoring of the analytical programme, Fig. 2.

Included are the controls for the zeroing of the detector bridge, the attenuation of the detector signal, the settings for and display of all the temperatures in the analytical unit and for the mode of operation, manual or automatic, for either channel.

The most important of the variable controls are the "attenuation $\times 1$", the "attentuation $\times 4$" and the "sample in delay" timers. The first two regulate the timing of the integral prints of the nitrogen and carbon dioxide peaks respectively; the last controls the introduction of the sample into the combustion zone relative to the slug of oxygen injected into the helium gas.

C. The integrator

This was specifically designed by Carlo Erba as an inexpensive, fully transistorized high precision integrator for use with their CHN analyzer. The use of an integrator operating on a time basis rather than one equipped with a slope detector or valley sensor is possible because the chromatogram from the analyzer contains at the most three peaks all of which are completely resolved. The main features of the integrator are shown in Fig. 4. The overall accuracy of this integrator is claimed to be $\pm 0{\cdot}1\%$.

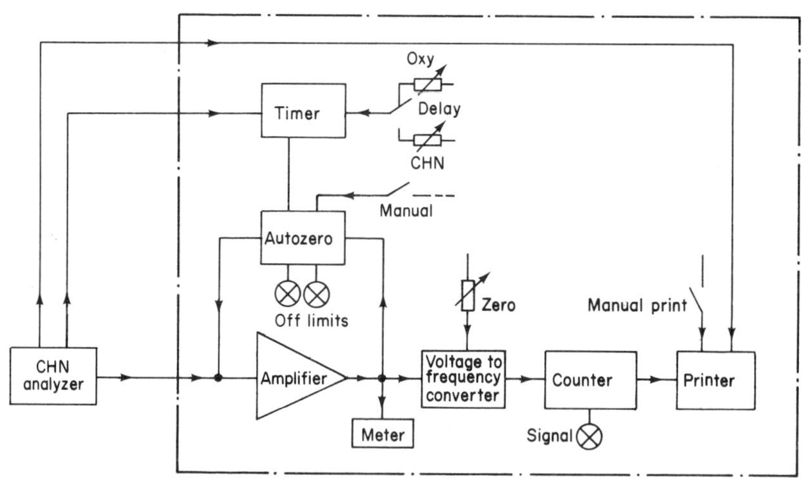

FIG. 4. Schematic integrator diagram.

The output of the CHN analyzer is coupled directly to the integrator. The signal from the detector is amplified in the servo amplifier and fed to a voltage-to-frequency converter which develops an output pulse rate (shown on the indicator signal lamp) directly proportional to the signal from the input amplifier. The output from the voltage to frequency converter operates a proportional counter which has a simple printing machine in parallel. At the beginning of each analysis, the integrator is automatically zeroed. The peak area integrals are printed at pre-set times. Whilst the timing of these prints is adjusted from the control unit, the automatic zeroing is set on the integrator itself.

Typical chromatograms and the sequence of analytical steps are shown in Figs. 5 and 6.

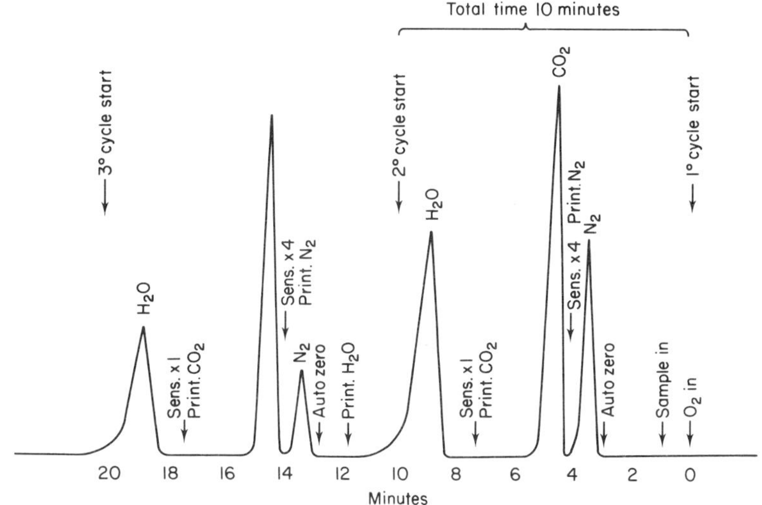

FIG. 5. Events sequence of CHN analysis.

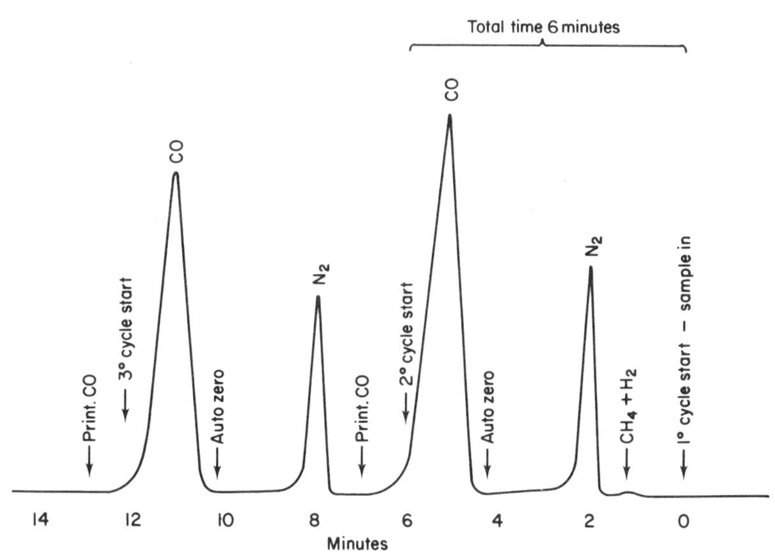

FIG. 6. Events sequence of oxygen analysis.

IV. PACKINGS AND FILLINGS: TUBES, FILTERS AND CHROMATOGRAPHIC COLUMNS

A. Combustion tubes and filter packings

Figures 7 to 9 show the filter packings, the latest combustion tube, and the reduction tube recommended by Carlo Erba. The key to the reagents used is Table 1.

Table 1. Key to reagents in Figs. 7–9.

A	– Chromium trioxide Cr_2O_3
B	– Copper silvered 10/1
C	– Cobalt(II), cobalt(III) oxide silvered; 3/1 Co_3O_4/Ag.
D	– Quartz wool
E	– Platinum wool
F	– Activated charcoal
G	– Activated Charcoal with Ni–Pt
H	– Mixture ascarite–anydrone 2/1
I	– Anhydrone or silica gel or sicapent
L	– Ascarite
M	– Anhydrone
N	– Copper sulphate · 5 H_2O
O	– Copper oxide
P	– Quartz turnings

Unless otherwise stated the reagents are supplied by Carlo Erba or are A.R. equivalents. The preparation and pretreatment of the special reagents are as follows:

1. *Preparation of the reduced silver-coated copper (Cu + Ag)*

Grind microanalysis grade copper oxide and sieve it to obtain a grain size between 0·4 and 0·7 mm. Add 7·5 ml of a saturated solution of silver nitrate to 100 g of CuO previously prepared. After water evaporation, reduce the dry mixture in a hydrogen stream at 500°C for 30 min and then at 700°C for 20 min. Then cool under hydrogen.

2. *Cobalt oxide, silvered*

Weigh 10 g of Co_3O_4 into an evaporating basin. Add a solution of 3·3 g A.R. silver nitrate in 3–4 ml of distilled water to the cobalt oxide and stir to a slurry. Heat the mixture gently until the decomposition of the silver nitrate is complete as shown by the disappearance of nitrous fumes.

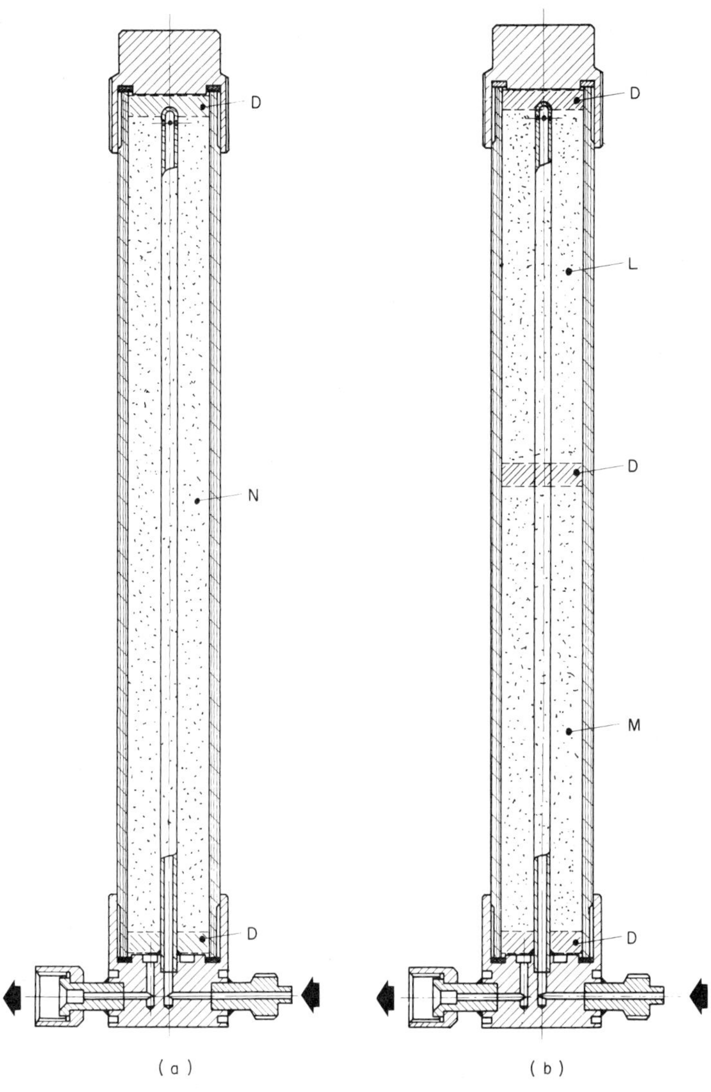

FIG. 7. A. CHN humidification tube; B. Filters for the auxiliary oxygen CHN channel and for the oxygen channel.

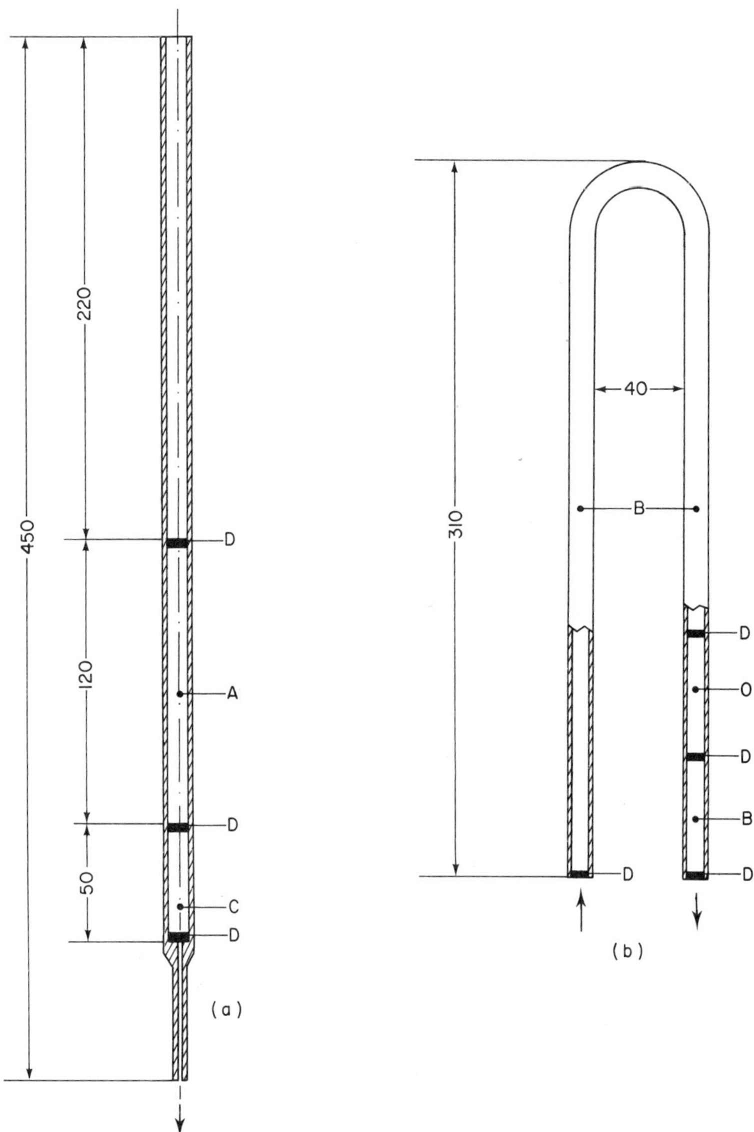

FIG. 8. A. CHN channel combustion tube; B. CHN channel reduction tube.

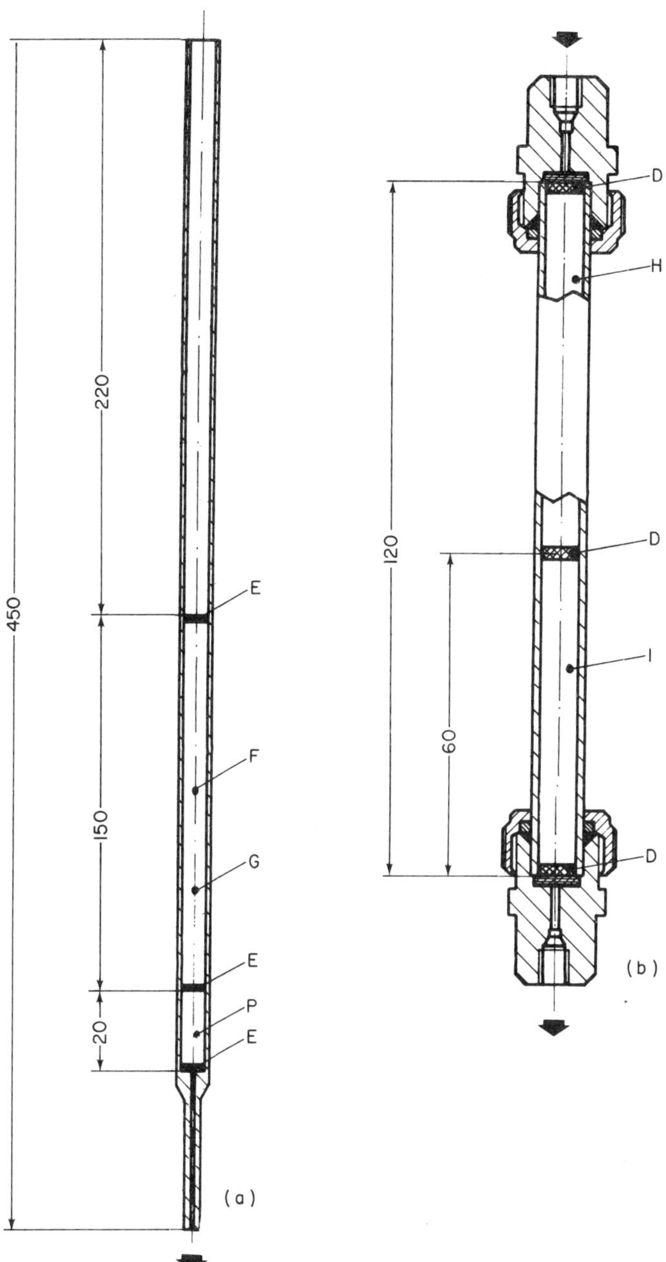

FIG. 9. A. Oxygen channel combustion tube; B. Oxygen channel filter for the removal of acidic gases

3. *Preparation of the activated charcoal*

Sieve the charcoal and collect the 30–80 mesh fraction. Extract the charcoal in a soxhlet with 10% hydrochloric acid for four hours. Wash with distilled water to remove the HCl. Dry at 100°C, and activate at 1000°C in a hydrogen stream (30 ml/min) for 12 hours until disappearance of hydrogen sulphide.

4. *Activated charcoal with Ni/Pt (special carbon CK3–Ni–Pt) patent pending* (Pella and Colombo[1])

Pelleted carbon, 25–120 mesh type CK3 (Degusa, Frankfurt, W. Germany) is heated at 1100°C for 8 hours in a hydrogen stream, then allowed to cool under hydrogen. It is then treated with a 10% solution of $Ni(NO_3)_2$ with a nickel/carbon ratio of 1:4 by weight. The slush is evaporated to dryness with stirring in an evaporating basin. The dry material is heated in a quartz tube to 150°C and flushed with hydrogen. The temperature is increased to 1000°C over 2 hrs. Between 300–500°C NH_3 is evolved; above 500°C nitrogen oxides are lost.

The material is heated at 1000°C for 30 min, then cooled to room temperature in a hydrogen stream.

The nickel-coated carbon is now treated with a concentrated solution of H_2PtCl_6 containing Pt in the ratio of 2:5 by weight with the nickel coated carbon. The mixture is rapidly evaporated to dryness in an evaporating basin with continuous stirring. It is then placed in a quartz tube at 200°C and flushed with hydrogen. The temperature is raised to 500°C and maintained for 30 mins before being increased to 1000°C for a further 30 mins. The reagent is cooled in a stream of hydrogen.

The resultant grey black bimetallic carbon should contain carbon, platinum and nickel in the ratio of 57·1:28·6:14·3 by weight.

5. *Copper oxide*

Lightly ground MAR copper oxide 20–30 mesh.

B. Chromatographic columns

1. *CHN channel*

The chromatographic column for the separation of nitrogen, carbon dioxide and water is 4 metres long, 0·5 cm i.d. and filled with Porapak Qs. The life of the Porapak Qs column is very long but when many fluorinated compounds are used its life may be greatly reduced.

2. *Oxygen channel*

The chromatographic column for the separation of carbon monoxide from the other pyrolysis products is 1 metre long, 0·5 cm i.d. and filled with molecular sieve 5 Å. If the molecular sieve 5 Å column efficiency should decrease due to moisture absorption, it should be reactivated by heating in an oven for about 2 hours, at 350°C in a stream of inert gas. This should always be done before using a new column.

The chromatographic columns can be used for several months and need only be replaced when the activity of a column is reduced and the peaks are flat.

V. PROCEDURE: PRE-TREATMENT, WEIGHING, LOADING, PURGING, COMBUSTION AND MEASUREMENT

After the instrument has been installed, and is leak free and thermally stable at the required temperatures, the flow rates in the CHN and the O channel are adjusted by the restrictor valves on the front of the instrument. The flow rates are adjusted to 30 ml/min ±1% in the CHN channel and to 25 ml/min±1% in the oxygen channels.

Before routine analyses several unweighed samples are burnt in each channel to ensure the optimum adjustment of the various timers. Details of this procedure are given in Section VIII.

A. CHN analysis

The tin sample containers are 4-mm diameter, 6 mm tall and weigh about 10 mg each. They are washed in acetone and dried at 140°–150°C before use.

1. *Weighing*

Approximately 0·6 to 1·0 mg of sample is accurately weighed into a tin capsule. For special analyses samples weighing 0·1 mg to 10 mg may be used. The container is closed with tweezers, the top being folded over twice, and care being taken not to spill any sample. The capsule is folded again to form a small pellet, care being taken to avoid splitting. This pelleting technique avoids any possibility of sticking in the magazine. The balance used is either a Sartorius 4125 or a Mettler 22, both of which have a digital display to 0·1 μg. The weighing must be as objective and strain free as possible if benefit is to be gained from the high output potential and precision of the instrument. Any balance fulfilling these criteria and capable of weighing rapidly to 0·1 μg is suitable.

2. *Magazine loading*

The magazine is loaded with an unweighed sample in position (1) to condition the combustion train at the start of each complete run of samples. Weighed standards are put in positions (2), (3), (4), (11), (17) and (23), nitrogen-free standards being used in positions (3) and (17) to provide a nitrogen blank under operating conditions. The apparently large number of standards is to ensure high precision and to check if any drift in factors occurs during a run. The magazine is put into the automatic sampler with position 0 above the combustion tube and the lid is closed with the screws provided.

3. *Purging*

The purge valve on the front of the sampler is slowly opened, left open for 30 seconds and then closed for 90 seconds. This is done 5 times. The valve is then opened for 2 minutes before thc purge valve is finally closed. This method of purging is preferred to that given in the manual because it gives a more stable baseline.

When the CHN pressure gauge on the instrument has regained its normal operating pressure, the attenuation is switched to "$\times 1$" and the recorder chart is started. When the baseline is stable, usually after 5 to 15 minutes, the bridge is balanced to give zero signal. There should be no signal from the integrator. If the instrument has been operated immediately before in the CHN mode, no switching is necessary. If it has just been used for oxygen analysis the switch must be changed from "oxygen" to "CHN" on the control unit and on the integrator. The "main" switch must be changed from manual to "AUTO ATT" and the cycle time reset to 10 minutes.

The attenuation is now switched to $\times 8$ and with the "sample-in delay" switch at off, the cycle is started. This ensures that no sample is dropped during the first cycle because the instrument design is such that the timing of peak prints in this cycle is several seconds earlier than in subsequent cycles. On completion of the cycle, the "sample-in delay" timer and "cycle" timer automatically reset, and it is when this occurs that the "sample-in delay" is switched to the "on" position. After a few seconds the "sample-in delay" timer starts and when it reaches zero a sample falls into the combustion zone. The combustion of the first sample is always watched at the start of a run to check that the oxygen injection and the sample introduction are correctly synchronized.

The introduction and combustion of samples and the recording of the corresponding integrals continue automatically until the cycle timer is switched off.

4. *Measurement and calculation of results*

The results are calculated by comparing the integrals for each element in the samples with those from the known standards. An electronic bench calculator is used to calculate the factors for carbon, hydrogen and nitrogen, and the mean values for the standards in a complete run are used to calculate the percentages of carbon, hydrogen and nitrogen in the samples of the same run. Peaks may be measured by area or height. The peak area is taken as a triangle and may be measured by planimetry, mechanical or electronic integration, or the product of peak height and width at one half height. Before any calculation the recorded peak values of the standard, the sample, and the blank run must be brought to the same sensitivity attenuation.

Apply the following formulae for the percentage calculation:

$$\%C = \frac{KCO_2\,(V_cCO_2\text{–}CO_2)}{P}$$

$$\%N = \frac{KN_2\,(V_cN_2\text{–}N_2)}{P}$$

$$\%H = \frac{KH_2O\,(V_cH_2O\text{–}H_2O)}{P}$$

where: N_2–CO_2–H_2O = blank value
K = response factor
V_c = sample peak value
P = sample weight in mg.

The response factor K for each element is determined as follows:

$$Ke = \frac{\%e \text{ theoretical standard} \times P_s}{V_s - V_b}$$

where: e = element to be determined
P_s = standard weight in mg
V_s = standard peak value
V_b = blank run peak value.

Factor K must be determined by averaging at least two runs for each series of analyses. More standards are preferable for better precision.

Example:

Standard weight: 1·052 mg of cyclohexanone-2,4-dinitrophenylhydrazone
theoretical %C = 51·79
theoretical %H = 5·07
theoretical %N = 20·14.

Sample weight: 1·125 mg of substance containing CHNOS
sample CO_2 peak area (S × 32) = 1100 mm^2
standard CO_2 peak area (S × 32) = 925 mm^2
blank run CO_2 peak area (S × 4) = 40 mm^2.

$$KCO_2 = \frac{51{\cdot}79 \times 1{\cdot}052}{925 - 5} = 0{\cdot}05922.$$

$$\%\text{C in sample: } \frac{0{\cdot}05922 \times (1100 - 5)}{1{\cdot}125} = 57{\cdot}64$$

sample H_2O peak area (S × 8) = 755 mm²
standard H_2O peak area (S × 8) = 810 mm²
blank run H_2O peak area (S × 8) = 8 mm².

$$KH_2O = \frac{5{\cdot}07 \times 1{\cdot}052}{810 - 8} = 0{\cdot}006650$$

$$\%\text{H in sample: } \frac{0{\cdot}006650 \times (755 - 8)}{1.125} = 4{\cdot}41$$

sample N_2 peak area (S × 8) = 200 mm²
standard N_2 peak area (S × 8) = 176 mm²
blank run N_2 peak area (S × 8) = 0 mm².

$$KN_2 = \frac{20{\cdot}14 \times 1{\cdot}059}{176} = 0{\cdot}12038$$

$$\%\text{N in sample: } \frac{0{\cdot}12038 \times 200}{1.125} = 21{\cdot}40.$$

B. Oxygen analysis

The capsules in this case are silver, each weighing approximately 20 mg. They are washed in acetone and dried at 140°–150°C.

The method is very similar to that for the CHN determination.

1. *Weighing*

Approximately 0·8–1·0 mg of sample (containing about 0·1 to 0·3 mg oxygen) is weighed into a silver capsule. This is folded to seal as for CHN and loaded into the magazine. The first two spaces are used for unweighed samples and standard samples are spaced at intervals, e.g. in spaces (3) (4) (10) (17) and (23) throughout the run.

2. *Loading and purging*

The magazine is loaded into the left-hand sampler above the oxygen pyrolysis tube, the top screwed down and the sampler purged as for CHN. The controls are set to the oxygen mode on the integrator and control panel, the "main" switch to "MAN ATT" and the cycle timer to 7 minutes. The attenuation is switched to $\times 1$, when the baseline is stable the bridge is balanced and the integrator checked for zero signal. The attenuation is set to $\times 8$ with the "sample-in delay" switch on and the cycle timer is switched on to "auto". No blank cycle is done at first, as for CHN, since there is no oxygen injection. The samples are dropped one by one into the pyrolysis tube and pyrolyzed, and the oxygen-containing products reduced to carbon monoxide.

3. *Measurement and calculation*

The corresponding carbon monoxide peak integral is recorded approximately $1\frac{1}{2}$ minutes into the next cycle. This time corresponds to the time of the water integral print when operated in the CHN mode. The calculations resemble those for CHN, an example being given

$$\%\text{O} = \frac{K\text{CO}\ (V_c\text{CO})}{P}$$

where: K = response factor,
V_c = sample peak value,
P = sample weight in mg.

The response factor for oxygen is determined as follows:

$$K = \frac{\%\ \text{oxygen theoretical standard} \times P_s}{V_s - V_b}$$

where: P_s = standard weight in mg,
V_s = standard peak value,
V_b = blank run peak value (V_b is normally zero for oxygen).

The factor K must be determined by averaging several values in each run.

Example:

Standard weight: 1·052 mg of cyclohexanone-2,4-dinitrophenylhydrazone
theoretical O% value = 23

Sample weight: 1·125 mg of substance containing CHONS
sample CO peak area = 490 mm²
standard CO peak area = 785 mm².

$$K\text{CO} = \frac{23 \times 1{\cdot}052}{785} = 0{\cdot}03082$$

$$\%\text{O in sample: } \frac{0{\cdot}03082 \times 490}{1{\cdot}125} = 13{\cdot}42.$$

VI. RESULTS AND SOURCES OF ERROR

A. Results

CHN. Our results with a wide range of organic compounds containing few interfering elements such as boron, fluorine, phosphorus and metals, have been excellent, being as good as or better than the classical methods. The standard deviations obtained by Stoffel(4) were C < 0·1, H < 0·015, N < 0·5, and other users confirm this. Early in our experience standard deviations were calculated from over 100 standards and gave values of 0·149 for carbon, 0·046 for hydrogen and 0·076 for nitrogen. Full details are reported in Section VIII. H.

B. Sources of error

The sample weighing, operator errors and sample homogeneity are all possible sources of error, particularly the latter in microanalysis, but since these are common to all micro methods, they will not be considered further. The errors which could be associated with the instrument are of particular interest.

1. *Peak area measurement*

The peak area measurement error can be minimized by following the proper method.

Table 2 shows the precision in absolute values obtained by Carlo Erba with different methods of peak evaluation.

It can be noted that, in some cases, the manual data have the same accuracy as the integrator data but, for the most part, while the response factor calculated on integrator data does not present errors higher than 0·2% for N, 0·3% for C and 0·15% for H_2, the manual calculations can present errors of 0·6%. For this reason the electronic integrator is recommended. Its accuracy is claimed to be 0·1% and unless the temperature conditions vary greatly, the results obtained give us no reason to doubt this. However, large temperature variations, e.g. 5°–6°C in an hour, can cause a large factor drift. We have two instruments, a Model 1102 and the later Model 1104, both of which are operated in an air-conditioned room controlled to ±1°C and this change in the integrator sensitivity only became apparent when the air conditioning failed. A steady drift was found in the factors calculated from the standards burnt during this temperature rise. On restoration of the original temperature, the factor resumed its former value.

A change in the temperature setting of an air-conditioned laboratory may

Table 2. The precision obtained by different methods of evaluation

Elements	Processing of data obtained from 20 analyses	Electronic integrator	Manual area	Manual height
C % Theoretical 51·79	Arithmetic mean	51·79	51·78	51·87
	Max. deviation	+0·29 −0·21	+0·42 −0·38	+0·6 −0·49
	Number of analyses with error >0·3%	—	6	9
	Standard deviation (σ)	0·18	0·25	0·3
H_2 % Theoretical 5·07	Arithmetic mean	5·07	5·07	5·07
	Max. deviation	+0·14 −0·17	+0·16 −0·2	+0·17 −0·18
	Number of analyses with error >0·3%	—	—	—
	Standard deviation (σ)	0·08	0·1	0·11
N_2 % Theoretical 20·14	Arithmetic mean	20·14	20·16	20·10
	Max. deviation	+0·18 −0·19	+0·36 −0·39	+0·35 −0·42
	Number of analyses with error >0·3%	—	3	6
	Standard deviation (σ)	0·12	0·22	0·25
O_2 % Theoretical 23	Arithmetic mean	23·00	23·02	22·95
	Max. deviation	+0·28 −0·20	+0·32 −0·25	+0·30 −0·27
	Number of analyses with error >0·3%	—	1	—
	Standard deviation (σ)	0·16	0·17	0·17
	Time needed for chromatogram quantitation of 20 analyses of CHN + O	—	1h 20 mins	30 mins

necessitate the fine adjustment of the integrator sensitivity setting. Full setting up details are given in the integrator manual and are not detailed here since they differ slightly for integrator Models 1110 and 1120. Another cause of factor drift is a sudden and marked change in atmospheric pressure. The latter is only a very occasional source of error.

2. *The quantitative conversion of the substance to carbon dioxide, nitrogen and water.*

This is governed by two factors, namely the combustion tube catalyst being maintained at the correct temperature, and the optimum combustion conditions being fulfilled. The latter depends upon the sample entering the combustion zone at the correct time relative to the oxygen enrichment of the carrier gas and is controlled by "the sample in delay timer". This is set so that the capsule burns with a marked "mushrooming combustion" starting about 3–5 seconds after the sample drops.

3. *The absence of gas leaks or blockages in the combustion train*

Should either of these conditions occur, the factors will become erratic. This is caused by the change in carrier gas flow rate through the train affecting the detector sensitivity. This change in flow rate also affects the synchronization of the sample drop and oxygen enrichment of the carrier gas. This is generally indicated by a "pre-glow" of the tin residues in the combustion tube before the sample falls, the prior passage of oxygen resulting in poorer combustion.

4. *Integral print timing*

The precision of the analyzer when operated automatically with an integrator, is dependent upon the correct setting of the integrator auto zero and on the printing of the peak integrals at the proper times.

VII. MAINTENANCE

A section on mechanical and electronic faults is hardly relevant since the instruments have been remarkably trouble-free with no recurrent faults. A list of faults which have occurred is included in Section VIII. L.

1. *Gas circuit filters*

Under normal operating conditions the filters are changed every six months, i.e. double the recommended interval, unless deterioration in the packing is evident.

The copper sulphate humidification tube is changed when about three quarters exhausted, usually after about three months.

2. *Tightness of O-rings*

The O-rings in the combustion tube and reduction tubes, samplers and connectors are changed whenever the tubes themselves are changed or emptied. Used O-rings are placed on top of the analytical unit and gradually resume their shape; they may be used again when no longer flat. If in any way damaged they are discarded to avoid any risk of gas leakage.

3. *Gas circuit leak check*

Although a daily leak check is advisable, it is omitted when maximum output is required since the flow rate control has to be closed, then opened and reset to the required value. Instead it is carried out when the tubes are changed.

Experience has shown that a slight fall in the pressure indicator for the CHN or O channel on the front of the instrument probably indicates a slight leak. Probably the best indication of a slight leak or blockage in either channel, even if not affecting the baseline stability, is the necessity to alter the bridge zero control. Little or no adjustment should be required from run to run or day to day, and this is confirmed by the consistency of factors. Alterations in the factor of 1% or more merit investigation except when any tubes have been changed.

4. *Combustion and reduction tube changing*

(a) *CHN channel.* The tubes may be changed without switching off the furnaces; hence this operation takes only a few minutes. When steel tubes packed with copper oxide and silver and operated at 950°C were used, they had a life of 600–700 analyses (3 weeks). Every 200–300 analyses (weekly) the capsule debris was emptied out. When the more recent tube packings in silica are used, they are replaced after 300–400 analyses (7–10 days). The Cr_2O_3 packing is practically indestructible and each packing serves for about 1000 analyses. The Co_3O_4 + 8% silver packing is changed every time the combustion tube is changed. The silica tubes are not re-usable.

The secondary reactor or reduction tube is changed after 600–700 analyses. This avoids gradual blocking of the reduction tube as oxidation occurs, and also avoids the tube becoming very difficult to empty.

After the tubes are changed the instrument may take up to 4 hours, usually less, to settle, so for convenience the tubes are normally changed at the end of a day.

(b) *The oxygen channel.* The life of the combustion tube depends on the type of compounds analyzed. Our experience has been limited but Pella and Colombo[(1)] suggest that their packing may be used for up to 1000 analyses.

5. *Auxiliary oxygen switching valve*

The oxygen switching valve can work for several months without replacement. When any trouble with the oxygen injection occurs the diaphragm should be replaced.

6. *Chromatographic columns*

These have already been discussed. Their life is long; only now, after three years, is our first CHN column showing any signs of deterioration.

VIII. ADDITIONAL INFORMATION

A. Initial setting up of the instrument

The instrument is installed, connected up and tested for the absence of leaks, and the flow rates in both channels are adjusted by the supplier. Comprehensive details for setting the instrument parameters are given in the instruction manual. However, a detailed description is given of the setting of the various timers, since correct adjustment is the key to the precision and accuracy of the instrument. Included is the timed sequence of analytical operations when operated in the automatic mode with an integrator.

1. *CHN channel—Detailed sequence of operations*

The magazine is loaded with several unweighed nitrogen containing samples and placed in the sample dispenser which is closed and purged (see Section V).

The "MODE" control is turned to the CHN position and the "ATTENUATION" control to "$\times 1$".

The "MAIN" switch is turned to the position "AUTO ATT".

The recorder pen is set to zero by the "ZERO" control and the baseline stability is checked.

The "ATTENUATION" control is switched to "infinity" and the "MANUAL" zero button on the integrator is pressed for 12 seconds.

The "ATTENUATION" control is switched back to "$\times 1$" and there should be no signal from the signal indicator lamp on the integrator.

The "AUTO ATTENUATION" timer controls are positioned as indicated in the test tables enclosed with the instrument and the "ATTENUATION" control is switched to "$\times 8$".

The "CYCLE TIME" timer is set to 10 minutes and the "SAMPLE-IN DELAY" timer is set according to the test tables.

For the first cycle the "sample-in delay" switch is left at "OFF" so that no sample falls into the combustion zone.

The "CYCLE TIMER" switch is set to "AUTO" and the run begins. The sequence in the first cycle differs in timing from subsequent cycles by the number of seconds delay between the "CYCLE" timers automatically resetting at the end of a cycle and their restarting for the next cycle. On our first instrument this is 28 seconds and on our second it is 15 seconds. At the end of the first cycle, when the timers reset, the "SAMPLE-IN DELAY" switch is moved to the "ON" position. The sequence is then as follows giving approximate times only since they vary slightly for each instrument.

Time 0. The oxygen loop (26) is inserted into the carrier gas circuit position "2" (Fig. 10) in series with the reaction tubes. (The cycle in delay timer starts. The auto attenuation timer ×1 starts.)

Time 1. 30 s. The oxygen switching valve is switched back to position "1" Fig. 10 and the loop is filled with oxygen.

Time 2. 25–35 s. Time preset on the delay timer, the drum advances one step and the sample falls into the reactor.
1 min 30 s. Water print from the previous analysis.
2 min 30 s. Auto zero light on the integrator comes on and remains on for 12 seconds.

Time 3. 3 min. The nitrogen peak is recorded.

Time 4. 3 min 15 s. Time preset on the timer "auto attenuation" ×1 stops. The nitrogen peak is over, and simultaneously the nitrogen integrator count is printed, the sensitivity is automatically switched and the ×4 lamp lights up and "attenuation" ×4 starts.

Time 5. 3 min 40 s. The CO_2 peak is recorded.

Time 6. 5 min. The CO_2 peak is completely eluted. Simultaneously the CO_2 integral is printed, the sensitivity is automatically switched back to ×1, lamp ×4 goes out and lamp ×1 lights up. The attenuation timer ×4 reaches zero.

Time 7. 6 min. The water peak is recorded.

Time 8. 10 min. The analytical cycle is complete and the next one starts. The water integral is printed approximately $1\frac{1}{2}$ minutes after the next cycle starts.

A detailed example for one of our instruments with the actual times recorded with a stop watch is given below. These times are used for the accurate setting of the instrument parameters as described below.

Action	*Time*
Cycle ends	0 s
Sample-in delay timer starts, oxygen loop switched in	15 s
Oxygen loop switched back	44 s
Sample falls and ×1 light on attenuator timer starts	51 s
Water prints from previous analysis	1 min 10 s
Auto zero light on	2 min 50 s
Auto zero light off	3 min 00 s

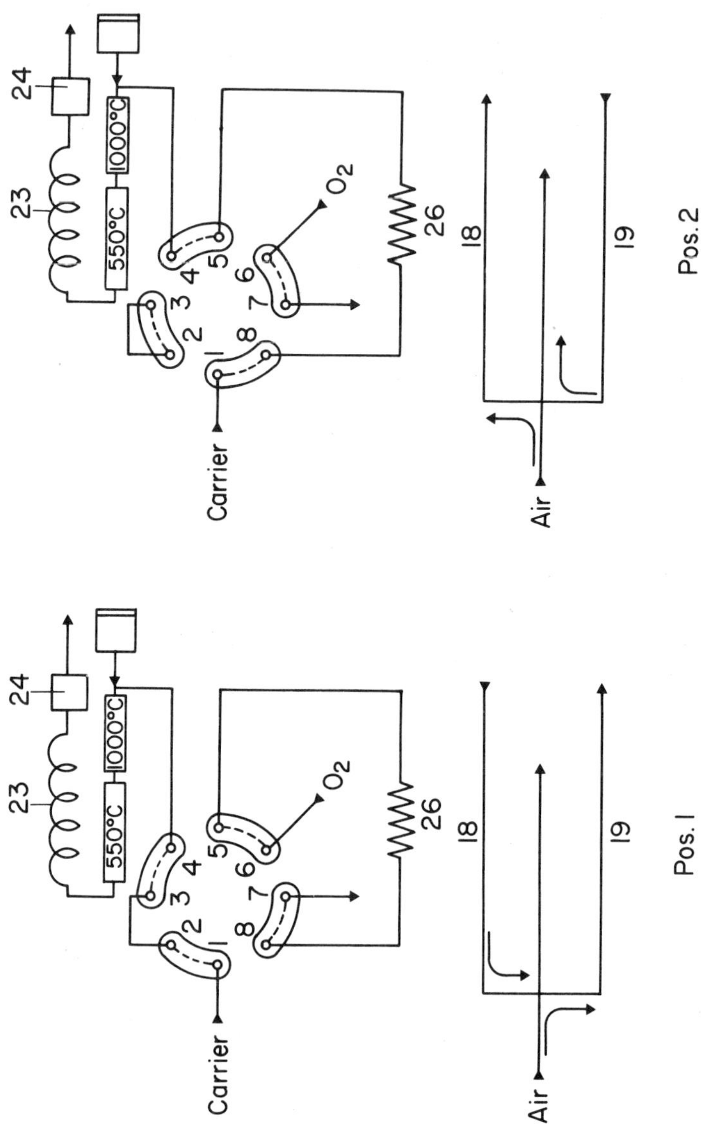

FIG. 10. Diagram showing oxygen injection sequence on CHN channel.

Action	*Time*
N_2 peak starts	3 min 31 s
N_2 peak stops	4 min 10 s
N_2 peak prints and ×4 light on and attenuator timer starts	4 min 29 s
CO_2 peak starts	4 min 46 s
CO_2 peak stops	5 min 43 s
CO_2 peak prints ×1 attenuator light back on	7 min 22 s
Water peak starts	7 min 53 s
End of run and timers reset	10 min 53 s

Timer Settings

(a) *Total analysis time.* If the water peak has not returned to zero before 10 minutes, the time set on the cycle timer is adjusted. This is rarely the case, and often the water peak is eluted so quickly that the cycle could be even shorter than 10 minutes.

(b) *Auto attenuation ×4.* The time set on the "auto attenuation" ×4 timer is not critical, since considerable time elapses between the CO_2 and H_2O peaks. If the timer is set to print about 30 seconds before the start of the water signal (as shown by the integrator signal light), plenty of margin is allowed for the CO_2 peak to be eluted should a sample be difficult to combust.

(c) *Auto attenuation ×1.* The ×1 to ×4 attenuation time should occur in the valley between the N_2 and CO_2 peaks. This is usually about 10 seconds before the CO_2 peak begins. This may be adjusted by using the auto attenuation ×1 times control.

(d) *The integrator auto zero* is set by the CHN delay control (screw driver) so that the auto zero light on the integrator comes on after the water peak has been recorded and goes out before the nitrogen signal begins (preferably about 15–20 seconds before) allowing plenty of time after the water peak has been printed.

(e) *Auxiliary oxygen–Sample-in delay.* The sample introduction should occur before the oxygen has reached a maximum in the primary reactor. This is timed to take place shortly after the beginning of the analytical cycle and is controlled by means of the auxiliary timer "sample-in delay". This timing is vital for optimum combustion conditions and should be checked together with the flow rate, whenever the carrier gas flow is changed for any reason, e.g. the replacement of the chromatographic column or the combustion or reduction tubes.

Where tin capsules are used, the optimum timing is readily seen by looking down the combustion tube through the transparent cover. The sample may flash on falling into the combustion zone; on coming to rest it should ignite and become incandescent. The best description is a "mushrooming" combustion becoming white hot as the oxygen reaches a maximum. The "mushrooming" should start 3–5 seconds after the sample falls and once seen is easily recognized.

A pronounced "pre-glow" of the tin residues from previous analyses before the sample drops means that the oxygen is arriving too early and the sample delay must be shortened or less efficient combustion will occur.

Our experience, particularly with copper oxide combustion tube packings, has been that correct timing makes the difference between good results and excellent ones. With the chromium trioxide packing, the effect is less marked but still appreciable, particularly with regard to the accuracy of the nitrogen results.

2. *Oxygen channel—Detailed sequence of operations*

The "Mode" control is turned to the "on" position and the sensitivity attenuation control is set to 1. The "main" switch is set to "MAN. ATT" and the "sample-in delay" timer level is set to position "ON". The recorder pen is set to zero and the baseline stability checked. With an integrator, the setting is then checked as for CHN, viz. with the attenuation at infinity the manual zero button is pressed for about 12 seconds, the attenuation is switched to ×1 and no signal should be shown by the integrator signal light. If any signal appears, this should be just cancelled by means of the "zero" control. After rechecking the baseline stability, the "cycle times" is set to 8 minutes and the sensitivity attenuation control "Attenuator" to "×8". The lever switch is placed in the position "AUTO START", the "sample-in delay" switch being left in position "ON".

The sequence is as follows:

Time 0. The analytical cycle starts.

1. (1 min): Time preset on the "sample-in delay" timer—the drum rotates 1 step and the first sample drops into the reactor.
2. (2 min 15 s): The hydrogen peak emerges.
3. (3 min): The nitrogen peak emerges.
   (5 min 15 s–5 min 30 s): Integrator auto zeroes.
4. (6 min): The CO peak emerges.
5. (6 min 30 s): The cycle is complete.
6. (8 min); The analysis is over.

(Note. Times 2, 3 and 4 may vary according to the efficiency of the chromatographic column.)

The only setting of importance, since there is no automatic attenuation, is the autozeroing of the integrator. This is set to light after the nitrogen ends and to go out about 15–20 seconds before the carbon monoxide signal commences.

The "sample-in delay" is normally left at the correct setting for the CHN analysis to avoid resetting. If this is altered after setting the integrator autozero, the latter must be altered correspondingly.

The detailed timings of an instrument follow:

Action	*Time*
Timers trip	0
Sample in delay starts	26 s
Sample in delay stops and sample drops	37 s
Carbon monoxide integral from previous sample prints	1 min 46 s
Nitrogen peak starts	2 min 12 s
Nitrogen peak ends	2 min 55 s
Auto zero starts	3 min 02 s
Autozero ends	3 min 12 s
Carbon monoxide peak starts	3 min 27 s
Cycle ends and timers trip and reset	7 min 52 s

B. Tin vs aluminium capsules for CHN

A throughput of nearly 1000 samples for CHN per month has meant that the instruments in the author's laboratory are always used in the automatic mode and the ideal of duplicate analyses except for special samples is precluded. For this reason an early study was made of the precision of the Carlo Erba with tin and aluminium capsules. The combustion tubes were stainless steel packed with copper oxide and silver wool and maintained at 950°C, the reduction tube temperature being 560°C. This study was carried out before the use of the copper sulphate humidification tube.

The data obtained for carbon, hydrogen and nitrogen values for three standard substances ignited in both aluminium and tin capsules were analysed.

For 108 standards, examined in 18 consecutive runs, in aluminium capsules, and 136 standards, examined in 23 consecutive runs in tin capsules, the results were shown by the χ^2-test to conform, or nearly conform, to Gaussian distributions for the three elements. Standard deviations are shown in Table 3.

Divergence from clear-cut Gaussian distribution can be attributed to the fact that the analyses were carried out in batches.

The F-test shows that marginally significantly better reproducibility is obtained for carbon and nitrogen results in tin capsules, but the reverse is true of hydrogen.

Table 3. Standard deviations with aluminium and tin capsules

Capsule	Standard deviation		
	Carbon	Hydrogen	Nitrogen
Aluminium	0·165	0·046	0·076
Tin	0·149	0·052	0·069
Overall	0·156	0·049	0·072

The standard substances which are readily combustible at the prescribed temperature of 950°C, are listed in Table 4.

With substances which are known to be difficult to burn efficiently, no true standards were available. However, their combustions proceeded far more efficiently and smoothly in the tin capsules. Results for carbon, hydrogen and nitrogen in nominally pure samples containing combined boron, phosphorus, potassium, sodium and fluorine as well as other halogens, sulphur and oxygen were almost always lower by as much as 1% relative in the aluminium capsules but in close agreement with expected values when tin capsules were employed. Results for hydrogen in compounds containing fluorine were detectably high but still analytically acceptable.

Table 4. Standard substances

Substance	Composition (%)		
	Carbon	Hydrogen	Nitrogen
2-Methylbenzimidazole	72·70	6·10	21·20
Phenacetin	67·02	7·31	7·82
Benzoic acid	68·84	4·95	—

On the whole, tin capsules are to be recommended because of reproducibility and of the better accuracy apparently shown in difficult samples. In addition, tin capsules facilitate the setting of the "sample-in delay" timer because of the enhanced combustion. Capsules are readily made from 0·025 mm thick tin sheet pressed in a sample die (Fig. 11). These tin capsules are approximately twice as thick as those supplied by Carlo Erba, because the die was originally designed for making aluminium capsules from 0·025 mm thick aluminium cooking foil.

FIG. 11. Capsule cutter and die.

C. Collaborative study

A collaborative study including an unknown compound in routine runs was carried out. The standard deviation for 10 analyses with an average weight of 0·82 mg of sample was calculated from $S = \{\Sigma(x - \bar{x})^2/(N-1)\}^{\frac{1}{2}}$. The sample was revealed to be *o*-chlorophenothiazine, and the results are shown in Table 5. Because of the long life of the steel tubes and the satisfactory results obtained, we have only recently changed to the $Cr_2O_3/Co_3O_4/Ag$ packings in silica tubes at 1050°.

Table 5. Results of collaborative study

	Theory	Found	S
Carbon (%)	61·526	61·55	0·150
Hydrogen (%)	3·421	3·51	0·045
Nitrogen (%)	5·981	5·91	0·043

D. Tailing peaks

The "signal" indicator bulb in the integrator as well as the chromatogram given by the recorder makes the observation of the start and finish of the signal peaks from a sample combustion a simple matter. Thus it is easy to see if a sample has been completely combusted by the absence of any tailing, particularly of the nitrogen and carbon dioxide peaks. There can of course be other causes of tailing and misshapen peaks, and many of these are given in the fault-finding manual with the instrument. This in fact is one of the reasons for using the chromatogram recorder even though the peak areas are not used for sample calculations.

E. Copper sulphate humidification tube

For the first two years, the Carlo Erba was operated in this laboratory without a copper sulphate pentahydrate tube. It was noticed that hydrogen values for compounds containing widely differing percentages of hydrogen, e.g. 1% or 11%, were poor. In addition the hydrogen factors found for standards with differing hydrogen values, e.g. benzoic acid (H = 4·95%) and phenacetin (H = 7·31%), were different, those for benzoic acid being the higher. A plot was made of the weight of hydrogen in a sample in micrograms against the digital peak integral obtained. This gave a reasonably straight line with a negative intercept on the *y* axis indicating that there was a loss of water. By adding a positive blank of this magnitude, i.e. of approximately 3·5 μg hydrogen equivalent to an integral count of 300 to each integral, the results became very good.

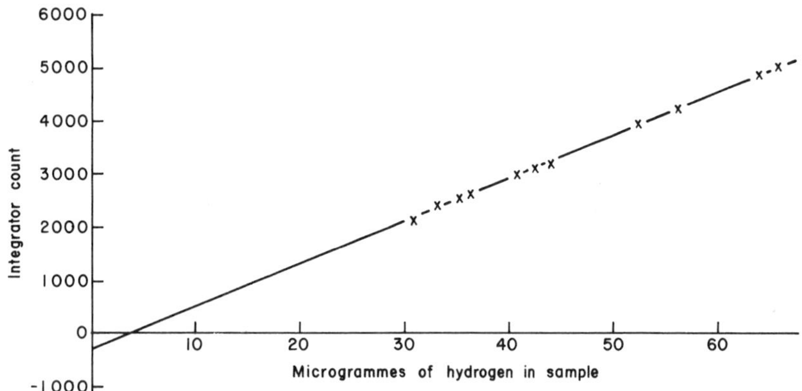

FIG. 12. Plot of weight of Hydrogen v. Integrater count showing Negative Intercept.

Since the $CuSO_4 \cdot 5H_2O$ humidifier tube was installed, the correction has become unnecessary and the water peaks are sharper. The humidification tube maintains a constant moisture level in the carrier gas, thus water absorption on the surfaces inside the CHN channel is minimized and a correction is unnecessary.

The earlier model Carlo Erba 1102 has a flexible coupling between the CHN combustion tube and the reduction tube. This consists of a PTFE tube surrounded by metal braid, itself insulated by Systoflex. The metal braid is soldered to eyelet type fittings which push into the brass couplings on the combustion and reduction tubes. This is to ensure efficient heat transfer to the metal braid, effectively heating the PTFE and preventing water vapour from the sample combustion condensing in the PTFE capillary. If the eyelets become loose, the heat transfer decreases, and the water peaks show signs of tailing particularly when high sample weights or samples with high hydrogen contents are combusted.

In the Model 1104 the flexible coupling is replaced with a silver tube. It is a good idea to wrap this connector with asbestos tape to improve the insulation and sharpness of the water peaks.

F. Blank values

In normal operation we apply only a nitrogen blank. The carbon blank is nil and the water blank small and negligible. The nitrogen blank is appreciable but reproducible to within $\pm 1\%$. It results partly from the air trapped in the capsule (this is very small) but mainly from the auxiliary oxygen supply. Its value, which may be up to 5 μg of nitrogen, only changes when the oxygen supply is changed.

The presence of a blank for nitrogen which is so reproducible as not to affect the results, is an advantage rather than a handicap in routine analyses of samples containing only 2% to 3% of nitrogen. The reason for this is because the lower the integral value, the more likely is the response of the integrator to be non-linear. This is shown by the following experience when, for a short time, the auxiliary oxygen supply was pure oxygen which gave no blank. Such oxygen is prohibitively expensive and wasteful, being bled into the atmosphere at 10–12 ml/min; and the results on nitrogen compounds with nitrogen values of less than 5% were lower than when medicinal oxygen was used as the auxiliary oxygen supply, because the nitrogen factor became non-linear at very low values. This may be overcome by standardizing the instrument with low nitrogen value compounds and grouping compounds accordingly. However, with a very high throughput this is inconvenient and it is much simpler if a wide range of compositions may be analyzed without "sorting".

The presence of the nitrogen blank aids in detection of the complete failure of the oxygen injection system since there will be no blanks on nitrogen-free samples and the other nitrogen values will be consistently low.

G. Computer calculation of results

Where facilities are available the use of a computer for processing data from the Carlo Erba is advantageous (Stoffel[(5)]). Both of our instruments have been fitted with tape punches in parallel to the integrator printing unit for use with an off-line computer. With one instrument the sample weights have to be separately typed on to punch tape but with the other instrument the system is being modified to incorporate a means of recording sample identification and weight directly from a small electronic memory.

The computer programme includes acceptability limits to the values for each factor, e.g. if a carbon factor value differs by more than 0·6% from the mean of the other values in a run, it is discarded. This excludes any obviously erroneous standards in a run which would grossly distort a factor. Similar limits are placed on the nitrogen and hydrogen factors.

In addition to calculating the mean nitrogen blank, factors, quotients and carbon, hydrogen and nitrogen percentages, the simplest empirical formula for each set of CHN data is given. The system has been in satisfactory operation for two years, the computer sheets forming a ready means of data storage.

H. Versatility

1. *Reduced sample size*

We have recently investigated the analysis for CHN of smaller samples than usual, i.e. for only 0·15 to 0·20 mg of sample compared with the usual quantity of 0·8 mg.

The four to five fold reduction in detector signal to the integrator consequent upon the decreased sample size is compensated for by altering the signal attenuation from ×8 to ×2. The resultant increase in sensitivity gives a four-fold increase in the nitrogen blank value. In addition very minor alterations to the instrument timing are necessary to allow for the increase in sensitivity. Three modified runs each containing 12 weighed samples (2-methylbenzimidazole, benzoic acid and phenacetin) were carried out. Three unweighed samples were included at the start of each run for instrument timing readjustment. The first run had a wide weight range from 0·18–0·40 mg (average 0·2837 mg), and the other two had ranges of 0·12–0·28 mg (average 0·1874 mg). There was little difference between the results, the higher weights being marginally the better. Under the modified conditions of reduced attenuation, acceptable results may be obtained on 0·15–0·2 mg. However, if low sample weights are included in a normal run, comparatively poor carbon figures are obtained whilst both the hydrogen and nitrogen values are low and unacceptable. This was demonstrated by including 11 weighed 2-methylbenzimidazole samples in the weight range 0·15–0·22 mg (average 0·1818 mg) in three separate runs. The mean results for these 11 samples were:

Carbon (%)	Hydrogen (%)	Nitrogen (%)	
72·70	6·10	21·19	Theory
72.69	5.64	20·74	Found

The standard deviations for the normal and modified methods were calculated according to the usual formula (p. 105) and the results were compared with the standard deviation for normal operation (see Table 6).

Table 6. Standard deviations for normal operation and for small samples

	Carbon (%)	Hydrogen (%)	Nitrogen (%)
Normal analysis	0·149	0·052	0·069
0·15–0·20 mg (analysis modified)	0·265	0·059	0·162
0·15–0·20 mg (normal run)	0·322	0·151	0·409

Thus to carry out analyses satisfactorily on sample weights of 0·15–0·20 mg, modified conditions are required and these small quantities cannot be done in routine operating conditions.

There is obviously wide scope, as has been reported in Carlo Erba bulletins, for varying the instrument parameters to cope with widely different samples particularly if the instrument is operated manually. Thus by using high weights, decreasing the attenuation for nitrogen and ignoring the carbon values, compounds containing only very small percentages of nitrogen may be analysed. Standards however should as far as possible be analyzed at the same attenuation settings.

2. *Difficulties in combusting compounds*

Attempts were made soon after buying our first instrument to combust carbon fibre samples using chromium trioxide at 1050°C but these were unsuccessful. No further attempts have been made. It is possible that slow burning compounds, e.g. carbon fibres, coal, etc., which are not readily pyrolyzed and combusted in oxygen by a dynamic system will always give trouble unless very small samples are used.

3. *CHN packings*

Whilst the normal recommended combustion tube packings have been given, it is highly probable that any packing which can be accommodated in the 18 cm (approx.) length of tube available and in the temperature range of up to 1080–1100°C can be used, provided that it is known to be suitable for a dynamic combustion system. In reality, the instrument provides a furnace with automatic sample introduction and oxygen injection and a means of separating and quantifying the products.

I. The analyses of liquids

One problem is the analysis of liquids. Involatile oils and gums can be analyzed normally, though difficulty may be encountered in weighing into the soft tin capsules. However, mobile liquids may escape and volatile liquids will almost certainly leak from crimped metal capillaries. The ideal method would be to encapsulate such a sample in a sealed glass capillary which cannot leak and will explode readily on dropping into the combustion zone.

The main difficulties are:

(a) making a glass container small enough to fit into the drum holes;
(b) obtaining consistent nitrogen blanks; and
(c) estimating an approximate weight of sample.

The first (a) was overcome by using small bulbs blown from melting point tubes. This method gave good carbon figures but (b) and (c) applied. The difficulty of estimating an approximate weight of sample forestalled the idea of flushing the bulbs with helium before filling, so another method was tried and proved successful.

Microcapillaries of standard bore, used in GLC work, of 25 μl capacity provided the answer. Each capillary is 64 mm long, i.e. 1 μl = 2·56 mm. The capillaries are cut into 8 mm lengths and one end is sealed using a fine oxyhydrogen flame. The capillary is weighed and the sealed end is inserted in a piece of fine bore PTFE tubing to facilitate handling. The open end of the capillary is inserted below the surface of the liquid to be analyzed, usually contained in a small specimen tube, and the whole is placed in a vacuum desiccator. The desiccator is fitted with a two-way tap, one way being connected to the vacuum and the other being open to air. By turning the tap to house vacuum and then to the air, the capillary is almost filled with liquid. If necessary it is tapped until only about 4 mm of liquid remains in the capillary and this is then centrifuged to the sealed end. The capillary is now held in a pair of forceps which have been precooled in Drikold and the open end is flash-sealed with a very fine oxyhydrogen flame. The sealed capillary, which is small enough to fit into the magazine of the Carlo Erba, is now reweighed. The blank values due to the small amount of air (1–2 μl) remaining in the capillary are virtually constant and negligible. If larger quantities are required wider micro capillaries may be used and the blank value calculated if necessary from the length of air remaining in the known bore of the capillary.

J. Unattended operation

The instrument can be left to run outside the normal working day and is switched off by shift technicians on the completion of a run. If this facility is not available a timed system for automatically switching off can be used, as has been developed by ICI colleague R. R. Sotheran *et al.*[6] They have also developed a method, based on a photoelectric cell, for automatically halting the sample drum if a container sticks in the combustion tube. This is a nuisance should it occur, since all the weighed samples following could be wasted.

K. Oxygen channel

Unsuccessful attempts were made to carry out direct oxygen analyses using "activated carbon" only. Even at 1140°C the oxygen response factor for benzoic acid was greater than the oxygen response factors for cyclohexanone-2,4-dinitrophenylhydrazone and *p*-nitroaniline. Replacement of the "activated carbon" by 50% Pt/carbon (Oita and Conway[7]) at 950°C gave consistent response factors for the three standards mentioned above but *lower* response factors for phenacetin and acetanilide. The oxygen response factor was given by

$$f = \frac{\%\ \text{oxygen theory} \times \text{standard weight } (\mu\text{g})}{\text{integral counts for CO}}$$

This "blank" occurring with phenacetin and acetanilide was presumed to be due to failure to separate the carbon monoxide from other pyrolysis products. At this time our attention was drawn to a paper by Pella and Colombo[1] in which 50% Pt/carbon[7] is said to cause the formation of large amounts of methane and the results are affected by temporary absorption of gases. Their paper, which extensively investigates the determination of oxygen in organic compounds by pyrolysis–gas chromatography, featured the use of the Carlo Erba Model 1102.

In their experience the method has the following drawbacks:

1. very high blanks arise from the pyrolysis of substances containing heteroelements, particularly chlorine, making the oxygen values unacceptable;

2. the absorption of nitrogen on the pelleted carbon interferes with the quantitative separation of CO from nitrogen, which would invalidate the gas chromatography method;
3. the diffusion phenomena in the pneumatic system reduce the conversion to CO and increase peak tailing.

However, these drawbacks will be found in all methods which use granular carbon prepared according to Unterzaucher[8] and which measure the product of the conversion, CO, or its product CO_2, by thermal conductivity.

In their study of the gas chromatographic determination of oxygen, to eliminate the drawbacks listed above, Pella and Colombo[1] modified the gas chromatographic unit, the reactor and the type of filling. The combustion tube is modified by capillary sections to eliminate dead space and the pyrolysis zone is maintained at 1100°C, about 100°C higher than the reaction zone, to ensure ready fusion of the silver containers and instantaneous pyrolysis (Fig. 12). A pure quartz crucible is used to collect the containers from about 200 analyses, otherwise the silver would form an alloy with the metals coating the carbon. The use of a 0·1 mm thick platinum cylinder in the pyrolysis zone is advised. As in all analytical systems using a gas chromatographic column, there is a pressure difference between the inlet and exit in the column. In order to increase this pressure, a capillary restriction (9) (Fig. 13) is inserted between the reactor and the scrubbing tube. This results in a positive pressure of 0·4–0·5 kg/cm^2 at the circuit inlet.

The nitrogen absorption on the carbon by the pyrolysis of nitrogen-containing compounds was studied in particular, and resulted in the use of the Ni/Pt-coated carbon described on p. 88.

In the case of the pyrolysis of nitrogen-containing compounds, Colombo and Pella[1] show how the oxygen/nitrogen ratio may be obtained from the ratio of the integrated values of the effluent CO and N_2 without weighing the sample used for analysis. They suggest that the integrated CO values, based on the CO/nitrogen value, may be correlated with the integrated values for CO_2, H_2O and N_2, obtained by combustion of the same sample in a different analytical circuit and measured with the same catharometer in the same carrier gas. Thus four integrated values proportional to carbon, hydrogen, nitrogen and oxygen may be obtained, by analyzing the sample, without weighing, in two different analytical circuits. Thus the atomic CHNO ratio may be found and hence the empirical formula for compounds containing only these four elements.

Attempts were made to use the Carlo Erba instrument in the author's laboratory with this Ni/Pt/carbon catalyst of Pella and Colombo. The only temperature recorded is that on the meter, and this was maintained at 1000°C. No extra capillary was inserted. Several runs were made with the standards shown in Table 7.

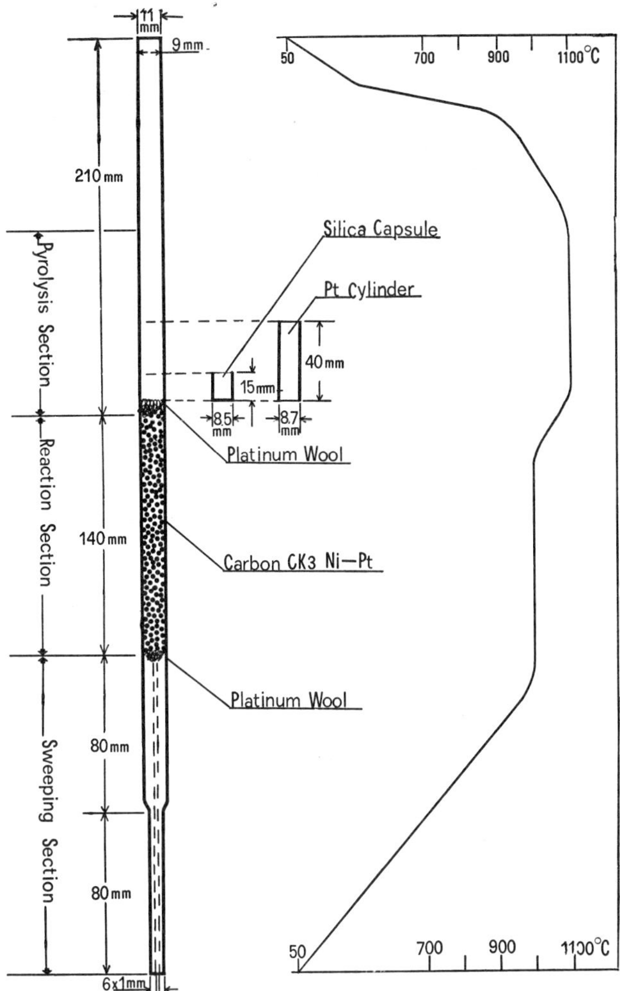

FIG. 13. Pyrolysis reactor, filling and temperature distribution for the oxygen channel (Pella and Colombo[1]).

The results omitting only those for *p*-chlorobenzoic acid which were invariably high, were statistically analyzed. Because of the widely differing percentages of oxygen, from 11·84% to 45·09%, in the standards, the error was expressed as a relative standard deviation, and was found to be 1·09%.

Blanks as determined by the inclusion of 2-methylbenzimidazole were virtually nil. It is concluded that compounds containing only C, H, N, S and O may be satisfactorily analyzed, but further work is required before

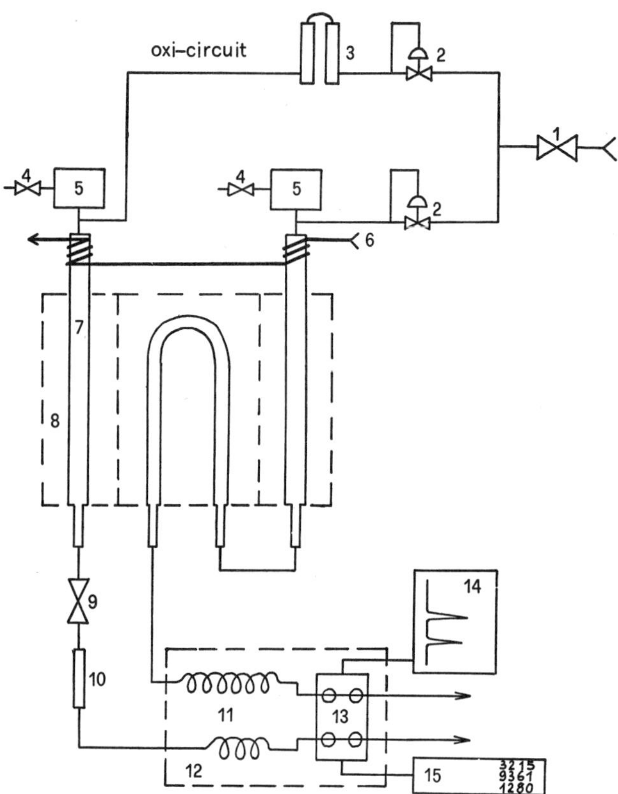

FIG. 14. Line diagram of apparatus.

(1) Pressure regulator
(2) Flow controller
(3) Purification tubes
(4) Sweeping valve
(5) Samples dispenser
(6) Water cooler
(7) OXY reactor
(8) Oven for OXY reactor
(9) Calibrated capillary tube
(10) Scrubbing tube
(11) Thermostated oven
(12) OXY chromatographic column
(13) Detector
(14) Recorder
(15) Digital integrator-printer

Table 7. Standards used

Standard	% Oxygen (Theory)
Cyclohexanone-2,4-dinitrophenyl-hydrazone	23·00
Benzoic acid	26·20
p-Nitro aniline	23·17
Phenacetin	17·85
Vanillin	31·55
Acetanilide	11·84
D-Glucose pentaacetate	45·09
Sulphonal	28·03
p-Chlorobenzoic acid	20·44

chlorine-containing compounds can be analyzed satisfactorily. Table 8 shows a typical set of results from a single run of analyses.

Table 8. Typical set of results from a single run

	%O Theory	%O Found	Error
Benzoic acid	26·20	26·31	+0·11
Benzoic acid	26·20	26·31	+0·11
p-Nitroaniline	23·17	23·14	−0·03
p-Nitroanaline	23·17	23·33	+0·16
Acetanilide	11·84	11·59	−0·25
Acetanilide	11·84	11·54	−0·30
Cyclobenzyl-2,4-DnPH	23·00	22·95	−0·05
Cyclobenzyl	23·00	22·84	−0·16
Phenacetin	17·85	17·75	−0·10
2-Methylbenzimidazole	Nil	Nil	—
Sulphonal	28·03	28·01	−0·02
Sulphonal	28·03	28·32	+0·29
D-Glucose pentaacetate	45·09	45·40	+0·31
D-Glucose pentaacetate	45·09	45·68	+0·59
p-Chlorobenzoic acid	20·44	22·35	+1·91
p-Chlorobenzoic acid	20·44	21·46	+1·02
Vanillin	31·55	31·73	+0·18
Vanillin	31·55	31·80	+0·25
Research sample	19·05	18·96	−0·09
Research sample	19·05	19·05	Nil

L. Electrical and mechanical faults

There have been no recurrent faults on either instrument in three years' usage, so only a list of possible faults and their effect is given.

1. Failure of auxiliary timer for oxygen resulting in no oxygen injection.
2. Failure of printer feed resulting in superimposed prints.
3. Failure of micro switch controlling the water print leading to the absence of the water integral.
4. Failure of a micro switch in the "$\times 1$" attenuator timer leading to absence of a nitrogen integral and a very large CO_2 integral due to the attenuation remaining at "$\times 1$".
5. Mechanical failure by fracture of the joint at the outlet of the reduction tube, indicated by a leak in the system.
6. Premature failure of the heater for the oxygen combustion tube through prolonged heating at 1140°C when first investigating direct oxygen analyses.

M. Omission of weighing

Reference should be made to a paper by Haberli(9) describing the use of the Carlo Erba (although this could be applied to any automatic CHN analyzer) where the fact is stressed that for simultaneous CHN determinations the sample weight need not be known thus resulting in considerable time saving. The arithmetic background for this procedure is presented, even for substances whose percentage of oxygen, chlorine and sulphur—if present at all—has been determined before. Based on this arithmetic procedure it is possible to calculate the empirical formula as well as the percentages of the elements present. Finally the programmes are briefly discussed which allow the simple calculation of the analyses—even by untrained members of staff—by means of a desk computer Olivetti Programme P-101.

CONCLUSION

The main disadvantage of the instrument is its inability to record an ash, e.g. a trace of silica as impurity in a sample which has been separated on a column and thus gives a low analytical result. A less serious one is that it is inconvenient, once a run has been started, to interrupt the cycle to insert an urgent analysis. This can usually be avoided by prior arrangement when urgent analyses are expected.

Its advantages, which greatly outweigh its disadvantages, are its high output, the ability to operate without attention outside the working day if required, simple routine maintenance, ease of operation and precision.

APPENDIX

Analysis of liquids on the Carlo Erba CHN analyzers

A great improvement in the sealing of the capillary tubes after centrifuging the sample to the bottom has been effected.

Whilst the part filling of a small weighed capillary with liquid under vacuum had been mastered, the sealing of the tube after centrifuging the sample to the bottom had presented problems. Previously the part filled capillary had been held in broad tipped tweezers precooled in Drikold whilst sealing with a fine blowpipe flame. The handling of these small tubes was difficult and sometimes the tweezers did not shield the entrained liquid sufficiently from heat to enable the tube to be sealed. The problem has been solved by using a small aluminium block in which a hole has been drilled just wide enough to take the capillary and 1 mm shallower than the length of the capillary. When inserted into the hole the protruding end of the capillary is readily sealed whilst the liquid it contains is fully shielded from heat. Where low boiling point liquids have to be analyzed the block itself may be cooled with Drikold.

Since this chapter was written Carlo Erba have introduced the Model 1106 Analyzer which can analyze for sulphur and/or oxygen as well as for CHN. No details have been included since the author's laboratory has had no practical experience of its use.

IX. REFERENCES

1. E. Pella and B. Colombo, *Anal. Chem.* **44** (1972), 1563.
2. E. Pella and B. Colombo, *Mikrochim. Acta* (Wien), (1973), 697.
3. R. Belcher, G. Ingram and J. R. Majer, *Mikrochim. Acta* (Wien), (1968), 418.
4. R. Stoffel, *Z. Anal. Chem.* **262** (1972), 266–268.
5. R. Stoffel, *Mikrochim. Acta* (Wien), (1972), 242–246.
6. R. R. Sotheran, Private communication.
7. J. Oita and H. S. Conway, *Anal. Chem.* **26** (1954), 600.
8. J. Unterzaucher, *Ber. Deut. Chem. Ges.* **73** (1940), 391.
9. E. Haberli, *Mikrochimica Acta (Wien)*, (1973), 597–606.

4. INSTRUMENTAL METHODS FOR THE DETERMINATION OF ORGANIC CARBON IN WATER

R. KÜBLER
Ciba–Geigy Limited,
Basle, Switzerland.

I. INTRODUCTION

The presence of solid and dissolved organic impurities in water has a strong influence on its use as potable water and general-purpose water. The organic impurities affect not only its appearance, smell, and taste, but also present an acute or latent hazard for people, animals, and plants. Furthermore, the oxygen deficiency caused by the biological and chemical degradation of these impurities endangers the living organisms required for its purification and recovery.

It is therefore of great importance to monitor the amount of organic impurities present in the water. This constitutes part of the efforts undertaken to prevent pollution of natural waters and decides ultimately on the need for purification steps and their success. The essence of this monitoring is to determine the multitude of organic substances as a total and, where possible to identify the individual substances or substance groups that differ by their harmful effect, and to draw conclusions with regard to the hazard created and the origin of the impurities.

This difficult task can be accomplished to a large extent by the use of modern methods available today. The efficiency of these methods is based on the optimum combination of preconcentration and separation steps with highly developed gas chromatography, mass spectroscopy and infrared spectroscopy. This approach is represented by the work of Grob:[(1)] in a single sample of sea water he was able to identify and to determine about 80 organic substances (most of which were not of biological origin) at concentrations ranging from the low parts-per-billion (ppb) to the parts-per-trillion (ppt) level. Although these methods are necessary, and provide useful information, they are also costly and laborious, and require expensive equipment and an experience and knowledge which are not always available. A further disadvantage of this approach is that the necessary extraction or preconcentration steps are not always quantitive. Thus, nonvolatile and water-soluble substances formed by the degradation of proteins and plant materials and/or originating from industrial waste waters, partly escape determination. For this reason great importance is attributed to the simpler methods which attempt to determine the organic substances on the basis of their reactions or common characteristics. These methods do not provide information with regard to the type and origin of the impurities. However, they offer a quick and reliable assessment of the degree of pollution and the efficiency of water treatment plants.

Measurement of UV absorption at 254 nm provides a rough indication for the degree of pollution of a water sample, for most of the substances present absorb more or less strongly at this wavelength. A good correlation between UV absorption and organic carbon content is however obtained only when the organic substances are present in high concentrations.[(2)]

Determination of "biological oxygen demand" (BOD_5) assesses the amount of organic substances present from their biological action. BOD_5 will detect all biodegradable compounds by measuring the oxygen consumed in five days but does not account for nondegradable or difficult-to-degrade materials.

The determination of "chemical oxygen demand" (COD) provides results in a much shorter time. The method determines all substances that can be degraded by wet-chemical oxidation with a dichromate/sulphuric acid mixture under standardized conditions. The oxygen required for the oxidation is calculated from the amount of dichromate consumed which is determined photometrically or potentiometrically. This method also gives incomplete recoveries with difficult-to-oxidize industrial waste waters. A further disadvantage of the method is that inorganic substances such as chloride ions also consume dichromate, thus leading to erroneous results in the presence of significant amounts of such substances as is frequently the case in industrial waste waters.

The "total oxygen demand" (TOD), i.e. the oxygen gas required for the oxidation of a water sample, can be determined for instance by the Ionics Model 225 TOD analyzer. The oxygen required for the combustion is taken from a helium–oxygen mixture supplied at a constant rate, the oxygen content of which is measured continuously by electrochemical means. The short drop in oxygen concentration recorded during the combustion is used to determine the amount of oxygen consumed. As in the determination of COD, the oxygen consumption also depends on the chemical character of the substances present and the presence of other elements such as nitrogen, sulphur and phosphorus. Thus, only a limited correlation exists between the values obtained and the carbon content of the water.

The importance of all these methods for the determination of BOD, COD and TOD is surpassed by the methods for the determination of *"total organic carbon" (TOC)*. The TOC is a reliable measure of the pollution of water with organic substances from the industry or from natural sources, since it embraces all dissolved organic substances and is not affected by the presence of other elements. Typical TOC values are shown in Table 1.

The following sections deal with the determination of TOC and the equipment used for this purpose.

Table 1. Typical organic carbon content of some selected water samples in mg/litre

Industrial waste water	5–10000
Factory water	2–3
River water (Rhine near Basel)	2–3
Potable water (Basel)	0·5–1
Redistilled water	0·1–0·2
Deionized water	0·1–0·8

II. DETERMINATION OF ORGANIC CARBON IN WATER

A. Decomposition

All available methods are based on the determination of organic carbon by complete oxidation of the organic substances present in the water and determination of the carbon dioxide formed. This oxidation can be achieved either by wet-chemical decomposition or by direct combustion at high temperatures.

1. *Wet-chemical oxidation.* Three basic methods can be used:

(a) Oxidation with a hot mixture of concentrated dichromate-sulphuric, iodic, and phosphoric acids;[3,4]
(b) Oxidation with a solution of peroxydisulphate in dilute sulphuric or nitric acids with silver nitrate added;[5,6,7]
(c) Oxidation in acidified solutions with oxygen gas bubbling through the solution and irradiation with UV light.[8,9,10]

The advantages of these methods are that large samples can be analyzed, low concentrations can be determined, and acids and salts do not interfere. The methods suffer from the fact that not all substances are oxidized quantitatively, that another combustion is frequently required to oxidize the incompletely oxidized volatile substances, that a relatively long time is required for a determination, and that carbon can be introduced with the reagents. Most of the methods are rather laborious and not particularly suitable for automation.

2. *The direct thermal combustion* of the water samples in combustion furnaces at temperatures up to 950°C in a stream of oxygen or with the use of oxidation catalysts can be applied successfully to samples with high carbon contents, so that only small amounts of sample from 20–2000 μl have to be combusted to obtain a satisfactory signal. The combustion methods achieve complete decomposition in a short time and are well suited to automation. Thus, most of the TOC analyzers available today are based on this oxidation principle.[11] A disadvantage of the method is that in some analyzers the activity of the combustion catalysts slowly decreases and that the combustion tubes are rapidly destroyed by water samples with high salt contents. A further disadvantage is that contamination of the injection syringe, sample inhomogeneity and errors in sample injection have a strong effect on the results, because of the small amounts of sample taken for analysis (usually less than 200 μl).

B. Determination of carbon dioxide

The methods used in TOC analysis for the determination of carbon dioxide are based on the following well-known procedures: titration in a nonaqueous solvent[12, 13] measurement of the electrical conductivity of alkaline absorption solutions,[14] measurement of IR absorption with a nondispersive IR gas analyzer,[15] reduction of carbon dioxide to methane and integration of the flame ionization detector (FID) signal,[16] retention of carbon dioxide on a molecular sieve column and integration of the thermal conductivity detector signal after heating of the column,[17] and measurement of the carbon dioxide concentration with a CO_2-specific electrode.[18] Of all these methods, the IR method is the most frequently used for TOC determination. It is based on the measurement of the IR absorption of carbon dioxide with a nondispersive IR gas detector, sensitive to CO_2. The detector is shown schematically in Fig. 1 and operates according to the following principle.

The modulated infrared radiation emitted by the source passes through two parallel gas cells of appropriate thickness, one of which is filled with a nonabsorbing gas, while the gas sample which contains the carbon dioxide flows through the other. The radiation emitted by the reference and sample cells hits the detector which consists of two chambers separated by a diaphragm capacitor and is filled with carbon dioxide which absorbs radiation of a wavelength of about 4·2 μm. When no carbon dioxide is present in the sample gas, the two detector chambers are heated equally by the absorbed radiation, and the pressure pulses of equal amplitude created by the modulation compensate each other. If the sample gas contains carbon dioxide, a fraction of the radiation of wavelength 4·2 μm proportional to the CO_2 concentration is absorbed in the light path through the sample cell. The two detector chambers are now heated differently, the pressure pulses between the reference and sample chambers of the detector do not compensate each other, and the capacity of the diaphragm capacitor changes. This capacity change is amplified and transformed into a signal which is proportional to the carbon dioxide concentration. Only large amounts of gases which have a strong absorption band in the immediate vicinity of the CO_2 absorption wavelength interfere in the determination.

The advantage of this purely physical method of measurement is its high sensitivity paired with high selectivity. The equipment is easy to operate and quickly prepared for operation, since it does not require chemicals, solutions, separation columns or auxiliary gases.

The other methods for the determination of carbon dioxide mentioned above will not be discussed further; they are described in sufficient detail in the literature cited.

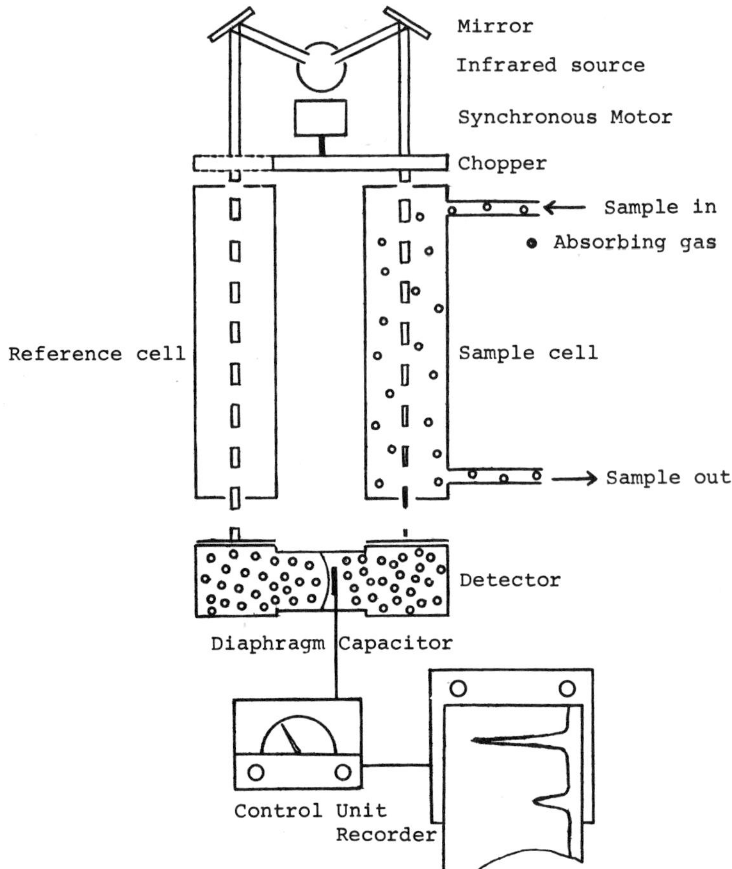

FIG. 1. Functional diagram of nondispersive Infrared Analyzer.

Besides the need for a quantitative decomposition and a reliable and sensitive carbon dioxide detector, the following factors, which will be discussed to more detail, have a decisive influence on the accuracy of the results:

The clear differentiation between inorganic and organic carbon.
Adequate sample preparation suited for a particular sample type.
Sampling, sample treatment and storage without contamination and losses.

C. Separation of inorganic carbon

The presence of inorganic carbon in most samples represents a basic problem in all TOC determinations. It is present in water as carbonate, bicarbonate or dissolved carbon dioxide and is liberated together with organic carbon as carbon dioxide in the oxidation of the sample. Some TOC analyzers provide a separate determination of this inorganic carbon (IC) which is then subtracted from the total carbon (TC) formed in the oxidation to obtain the TOC value:

$$\text{TOC} = \text{TC} - \text{IC}.$$

This *difference method* is applied mainly to industrial waste waters with high organic carbon contents. However, at carbon contents of less than 5 mg/litre the TOC values obtained by difference are unreliable, since neither the TC nor the IC values can be determined with sufficient accuracy. This is the case particularly when the inorganic carbon content exceeds the organic carbon content.

The *purging or degassing method* is recommended in the case of water samples with an organic carbon content of less than 5 mg/litre; the inorganic carbon is removed before the TOC determination by acidification of the sample to pH 1–2 and removal of the carbon dioxide liberated with a CO_2-free carrier gas. The inorganic carbon content of river water can exceed the organic carbon content by a factor of 20 to 50. For this reason, the IC removal must be carried out with special care, since the accuracy of the values and the detection limit for organic carbon depends on the completeness of CO_2 removal.

Table 2. Changes in the organic carbon content of aqueous solutions of organic substances after purging with nitrogen. (Flow rate 50 ml/min, purging time 5 min, amount of solution taken 3 ml.)

Substance dissolved in CO_2-free deionized water, acidified to pH = 1	Carbon content before purging (mg/litre)	Carbon content after purging (mg/litre)
Water, deionized	0·2	0·2
Acetone	6·0	5·3
Methanol	3·3	3·3
Isopropanol	4·8	4·7
Tetrachloroethane	2·7	0·6
Isooctane	5·2	0·2
Tetradecane	0·8	<0·2
Motor gasoline	4·9	<0·2
Fuel oil	2·4	0·5
Benzene	5·2	<0·2
Toluene	4·6	<0·2
Naphthalene	2·5	0·6
Potassium hydrogen phthalate	6·3	6·2

A disadvantage of the degassing method is that volatile organic constituents will be removed from the water together with the carbon dioxide. Table 2 lists the volatilities of some selected substances.

In the case of industrial waste waters with high solvent contents, this loss leads to large errors. It can be avoided to a large extent by trapping of the volatile substances in cold traps or on separation columns and subsequent separate combustion of the trapped substances. If the sample contains only traces of volatile organic substances, their quantitative determination and separation from carbon dioxide is difficult and subject to losses. The trapping of volatile substances is therefore omitted when purging carbon dioxide from surface waters, based on the assumption that the volatile substances originally present have already evaporated due to the intensive contact of the water with the atmosphere. The analyst must decide in each particular case to what extent this assumption is justified.

D. Sample preparation

A prerequisite for a correct and reproducible TOC determination is a homogeneous water sample, since the organic carbon can be present in the water in soluble form as well as in the form of suspended particles. It is evident, particularly in the case of small samples of less than 200 μl, that such particles can be the cause of gross errors. If the sample is cloudy, if solid particles are suspended in it, or if a sediment forms on the walls and the bottom of the sample container, the sample must be homogenized as much as possible by shaking, stirring and dispersing.

If this is not achieved and the sample remains a cloudy suspension containing particles of different size which settle at different rates in the sample container, the syringe, etc., the sample must be filtered or centrifuged. In such cases only the "dissolved organic carbon" (DOC) can be reported, neglecting the suspended particles. Membrane filters with a pore size of 0·45 μm are used for the filtration and solid particles that pass this filter are generally considered as dissolved.

E. Sampling and sample storage

The use of sensitive equipment is justified, and satisfactory results can be expected, only if the analysis is combined with adequate sampling, sample storage, and pretreatment. The universal presence of organic substances in the form of dust, detergents, fats and evaporation residues makes it essential for sample containers, filtering devices, transfer pipettes, syringes, etc., to be cleaned thoroughly before use. Equipment made of organic materials instead of glass should be avoided and flexible tubing may be used for sampling only where a strong and constant sample flow keeps the concentration of substances released by the tubing material at a low level. The distilled or deionized water used for cleaning itself contains some organic substances. The best cleaning method is therefore to use the water sample which is usually available in abundant quantities, as the cleaning medium, and to flush repeatedly the sample containers, sample transfer vessels and syringes used. The sample container should be as large as possible (0·25–1 litre) and should be filled completely to achieve a favourable ratio of sample volume to glass surface.

One of the commonest sources of error is instability of the sample. Organic substances can precipitate out during storage, become adsorbed on the container walls, separate out on the surface, or migrate into the vapour space above the liquid. Transfer of the sample from one container to another should be avoided, since some substance can be lost or introduced in this operation. Biological degradation during storage can be prevented efficiently by acidification and freezing of the sample. The inorganic carbon content must remain the same if the TOC is to be determined by the difference method, at least for the duration of the analysis. Industrial waste waters can be basic as well as acidic; it is therefore necessary to consider that carbon dioxide may be picked up or released and to prevent this by appropriate measures.

III. INSTRUMENTS FOR THE DETERMINATION OF TOC

Numerous methods have been developed for the determination of carbon dioxide in water. The most successful of these methods provide the basis for the automatic TOC analyzers which are now available commercially. Some of these analyzers are listed in Table 3. The principles of operation, their main applications and characteristics are described in the following sections.

In the first instance, this compilation provides a survey of the principles of TOC determination on which the commercial equipment is based and is not intended as a comprehensive list of such equipment. The selection is restricted to analyzers suitable for the analysis of discrete samples and does not include TOC monitors designed for the continuous survey of water treatment plants or numerous TOC procedures described in the literature which are not yet available commercially.

The order of presentation is not related to the quality or significance of the analyzers. It frequently depends less on the analyzer itself than on the experience of the analyst and his confidence in the measuring principles of the analyzer whether a given problem can be solved with a given analyzer. The microanalyst will use different equipment and techniques from the gas

Table 3. Instruments for the determination of TOC.

Instrument	Oxidation method combustion temperature	Method of deter- mination of carbon dioxide	Method of TOC determinati 1. Difference method (diff) 2. Oxidation after removal o inorganic carbon by purgi acidified sample (purg)
Beckman Model 915-A TOC analyzer	combustion in air cobalt oxide catalyst 950°C	nondispersive IR Analyzer	diff
			purg
Ionics Labor TOC Analyzer 445	combustion in N_2 oxidation catalyst 950°C	nondispersive IR Analyzer	diff
W. C. Heraeus Merz Rapid C TOC analyzer	combustion in O_2 copper oxide catalyst 900°C	nonaqueous titration	diff
			purg
H. Maihak TOC–UNOR	combustion in air oxidation catalyst 850°C	nondispersive IR Analyzer	purg
Carlo Erba Model 420 total carbon monitor	combustion in N_2 copper oxide catalyst 950°C	conversion to methane, integration of FID signal	purg
Dohrmann DC 52 TOC Analyzer	pyrolysis in He cobalt oxide catalyst 850°C	conversion to methane, integration of FID signal	purg (at 110°C)
			purg
Oceanography International total carbon system	wet oxidation potassium peroxydisulphate 175°C	nondispersive IR Analyzer	purg or diff
	combustion in O_2 950°C		purg
Wösthoff Hydromat–TOC	wet oxidation dichromate/sulphuric acid 160°C	conductivity measurement of absorbing solution	purg
H. Maihak UV–DOC–UNOR	wet oxidation with O_2 irradiation with UV-light 40°C	nondispersive IR Analyzer	purg

)etermination of 'olatile organic arbon possible	Required sample volume	Ranges (mg C/litre)	Repeatability	Analysis time per sample
no	20 µl	0–4000	±2% of scale	2–4 min
no	200 µl	0–5	±5% of scale	—
no	50 µl	0–20 0–3000	±2% of scale	3–4 min
no	0·5–5 ml	0–2000	±0·5% at >50 mg/l	8 min
no		0–200		
yes	continuous addition 20–80 ml/h	0–10 0–100 0–1000	±3%	20 min
yes	20 µl	0–25 0–250 0–2500	2% of scale	5 min
yes	30 µl 10 µl	0–2000 0–6000	1 mg/l or ±3% 3 mg/l or ±3%	5 min
no	10 µl 100 µl	0–200 0–20	0·5 mg/l or ±3% 0·05 mg/l or ±3%	5 min
no	10 ml	0–10	±2%	8 h oxidation time
no	100 µl	10–50000		2–3 min
no	continuous addition by proportioning pump	0–100 0–500	±2% ±5%	20 min 6 samples/h
yes	0·05–20 ml	0–0·1 0–10	±2%	3–12 min

chromatographer, the spectroscopist or an inexperienced equipment user. Frequently, it is this special experience and knowledge of a particular method that leads to successful use of the equipment.

A. Beckman, Model 915-A total organic carbon analyzer

The Beckman TOC Analyzer and the Ionics 445 Laboratory TOC Analyzer are based on the oxidation of the sample in the gas phase and IR spectroscopic determination of the carbon dioxide formed. This method, originally developed by Van Hall and Stenger,[11,15,19] which corresponds to the ASTM Method D-2579-69, has been the source of numerous new designs. Its principle (in modified form) also provides the basis for other TOC analyzers. A detailed description of the method follows, therefore, based on the design and operation of the Beckman TOC analyzer.

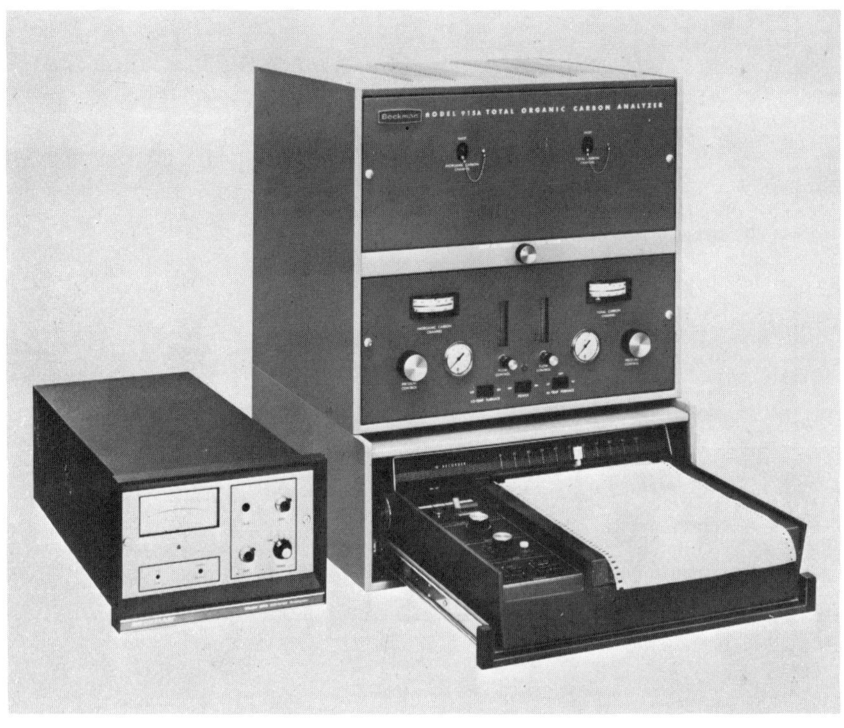

FIG. 2. Beckman, Model 915-A. Total Organic Carbon Analyzer.

1. *Principle of the method*

The TOC content is determined by injecting 20 μl of the water sample into a combustion tube heated to 900–950°C. The organic substances present in the water are completely converted to carbon dioxide with the aid of an oxidation catalyst and the oxygen-containing carrier gas. The gas stream leaving the combustion tube is passed through a water separator to remove the water vapour and led into a nondispersive infrared carbon dioxide analyzer, which gives a continuous record of the CO_2 content of the carrier gas. Inorganic carbon, which is also liberated under these conditions, is included in the measured value, so that the height of the peak appearing on the recorder chart is proportional to the total carbon (TC) present in the sample.

Another 20 μl of the same sample is then injected into another reaction tube which is heated to 150°C and packed with quartz chips, wetted with phosphoric acid. The acid in the tube liberates carbon dioxide from the carbonates present in the water, which is then, after removal of the water vapour, flushed by the carrier gas into the IR analyzer. The height of the signal recorded is proportional to the inorganic carbon content (IC) of the water sample. The total organic carbon content (TOC) is then obtained from the difference between total carbon (TC) and inorganic carbon (IC):

$$TOC = TC - IC$$

2. *Equipment*

The design and operation of the analyzer are shown by the scheme in Fig. 3. The furnace module contains two separate, practically identical

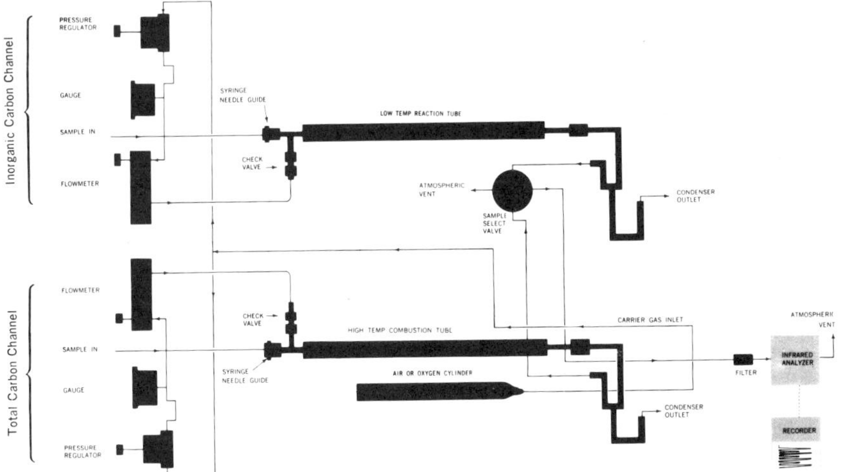

FIG. 3. Beckman, Model 915-A. TOC-Analyzer. Flow diagram.

channels. The first channel with the high temperature furnace serves for the determination of total carbon (TC), the second channel with the low-temperature furnace for the determination of inorganic carbon. Oxygen or air supplied from the same pressure cylinder passes through both channels and serves both as carrier and reactant gas. The gas pressures and flow rates can be controlled and monitored separately for each channel by means of two sets of flow and pressure regulators, manometers and flow meters. The constant gas stream passes through an injection block where it picks up the amount of water injected and carries it to the combustion and reaction tube, respectively. The water vapour leaving the furnaces is condensed in the corresponding condensers and the water-free gas proceeds through a four-port valve and a filter to the IR analyzer.

The four-port valve serves to connect alternatively the TC or the IC channel to the IR analyzer and to discard the gas from the unused channel to the atmosphere. The two channels differ in the respective packings of the two reaction tubes and their operating temperatures which can be adjusted independently.

The combustion tube of the TC channel is made of Hastelloy X or quartz and packed with 5 to 6 cm of cobalt oxide on asbestos according to the manufacturer's instructions. It is kept at a temperature of 900–950°C.

The low-temperature reaction tube in the IC channel is made of Pyrex glass and packed with 10 cm of quartz chips wetted with 85% phosphoric acid. The furnace temperature is 150–180°C and is high enough to accelerate the purging of water vapour and liberation of carbon dioxide, but not high enough to cause oxidation of organic substances.

The carbon dioxide concentration in the oxygen-carrier gas mixture is measured with the Beckman Model 865 IR Analyzer and displayed on a recorder. This IR analyzer, based on the nondispersive principle with the ranges 0–100 ppm and 0–500 ppm CO_2, has a sensitivity and reproducibility sufficient to detect 0·1 mg carbon per litre of water in a 20-μl sample. The carbon dioxide formed in the combustion passes through the analyzer in a few seconds so that the height of the peak displayed on the recorder is directly proportional to the carbon content of the sample. Integration of the very narrow CO_2 peak is therefore not necessary.

3. *Reagents and standards*

Standard combustion tube (Beckman Instruments) Packing: cobalt oxide on asbestos.

Low-temperature reaction tube (Beckman Instruments) Packing: phosphoric acid on quartz chips.

Reactant gas: Air or oxygen from pressure cylinders, free of carbon dioxide and hydrocarbons.

Water, free of carbon dioxide, for the preparation of standard solutions and for dilution: add 1 ml concentrated hydrochloric acid (analytical reagent grade) to 1 litre of distilled water and purge with nitrogen for 15 min.

Standard solution, 1000 mgC/litre, for calibration of the TC channel: dissolve 2·125 g of potassium hydrogen phthalate (Reagent Grade) or 5·665 g of sodium acetate trihydrate (Reagent-Grade) in 1 litre of distilled water.

Prepare by dilution standard solutions containing 10, 20, 40, 60, 80, 100, 200, 400, 600 and 800 mgC/litre.

Standard solution, 1000 mgC/litre, for calibration of the IC channel: dissolve 4·404 g of sodium carbonate (analytical reagent grade) and 3·489 g of sodium bicarbonate (analytical reagent grade) in 1 litre of distilled water. Prepare by dilution standard solutions containing 10, 20, 40, 80, 100, 150, 200 and 250 mg/litre.

The diluted standards are prepared by diluting aliquots of the concentrated stock solution with carbon dioxide-free distilled water, with the use of glassware cleaned with chromic acid. The diluted standard solutions should be kept in the refrigerator to improve their stability.

4. *Procedure*

Switch on the IR analyzer, the recorder and the furnace module. They are ready for operation after a two-hour heating period. The temperature of the TC furnace should be 900–950°C, and the temperature of the IC furnace 150–180°C. Freshly packed combustion and reaction tubes must be conditioned for at least 12 hours under operating conditions (with the IR analyzer disconnected) to obtain a drift-free and stable baseline. For the same reason and to protect the combustion tube it is recommended to leave instruments which are used every day switched on overnight. The flow rate of oxygen or air is adjusted by means of the flow controllers to 100 ml/min for both channels. It can be reduced to 10 ml/min when the equipment is not in use. The gain of the IR analyzer is chosen so that a full-scale deflection is obtained when 20 μl of a 100-mgC/litre standard is injected. Under these conditions the noise level of the signal should not exceed 0·5% of the full-scale deflection. In the case of a higher noise level, all gas lines leading to the IR analyzer must be inspected and cleaned. The instrument operation is considered as satisfactory when repeated injections of the same standard produce the same signals that return to the baseline in a few seconds.

Both channels must be calibrated with the appropriate standard solutions before a TOC determination is carried out. For this purpose 20-μl portions of the diluted standards prepared from the stock solutions are injected into the channel to be calibrated and the peak heights obtained plotted against the carbon content of the standard solutions. The carbon contents of the samples are then read from the calibration plot. The gain of the IR analyzer is adjusted so that a peak height representing about 90% of the full deflection is obtained when 20 μl of the standard with the highest carbon content is injected.

The standard solutions are injected with the simple and reliable 50-μl spring-loaded Hamilton CR-700-50 syringe. The syringe is flushed several times with the standard solution and then filled, care being taken that no air bubbles are trapped in the syringe barrel. The injection needle is wiped and the loaded syringe inserted into the injection block; the sample is injected and the syringe withdrawn when the signal on the recorder has returned to the baseline. To increase the reliability of the results, it is recommended to carry out duplicate determinations and to interpolate blank determinations at regular intervals using high-purity water.

The carbon content of the water samples to be analyzed is determined in the same way as that of the standard solutions, keeping in mind the precautions mentioned in Sections II.D and II.E. It may be necessary, therefore, according to their condition, to homogenize, filter or, in the case of alkaline industrial waste waters, to acidify the samples to achieve reproducible results.

To carry out a determination, 20 μl of the water sample is injected in the TC channel. After appearance of the peak on the recorder, the four-port valve is switched to the other position and the same amount of sample injected in the IC channel.

The peak heights obtained are measured and the TC and IC values of the water samples obtained from the calibration plot. The TOC content of a sample is obtained by subtracting the average value of a duplicate IC determination from the average value of a duplicate TC determination, taking into account the dilution factor V if the sample was diluted prior to determination.

$$\mathrm{TOC} = (\mathrm{TC} - \mathrm{IC}) \,.\, V \qquad (\mathrm{mg\,C/litre})$$

Should the carbon content of the sample exceed the calibration range, it must be properly diluted with carbon dioxide-free water. A sample dilution is also recommended if the sample contains more than 5% of acids or salts. In the case of carbon contents of less than 10 mg/litre the amount of sample injected may be increased but should not exceed 200 μl, as the large amount of water vapour formed affects the pressure in the tubes and may cause incomplete combustion.

The precision of the difference method is not sufficient for the determination of the TOC content of surface and drinking water (see Section II.C). The inorganic carbon must therefore be removed from the sample by purging. For this purpose, about 20 ml of sample is transferred to a suitable container, adjusted to pH 2 with concentrated hydrochloric acid and the liberated carbon dioxide removed from the sample by purging for five minutes at a gas flow rate of 50–100 ml/min. Two hundred microlitres of the sample treated in this way are then injected into the TC channel of the analyzer. The inorganic carbon has already been removed from the sample, hence the peak height measured corresponds directly to the TOC content of the sample.

5. *Results and sources of error*

The results obtained by the difference method described show that the analyzer should be applied mainly in the range 5–4000 mg C/litre. The reproducibility of the TOC values in the range 50–4000 mg C/litre is about ±2% of the range chosen and, in the case of TOC values around 5 mg C/litre on a range of 5 mg, about ±5% of the full recorder pen deflection. However, when the inorganic carbon predominates, the scattering of the TOC values is higher because it is obtained by difference and because of the instability of acidic solutions. Under normal operating conditions the limit of detection in the case of surface waters with an inorganic carbon content of up 50 mg/litre is 5 mg/litre, i.e. a multiple of the sensitivity attainable by the IR analyzer. If the presence of volatile compounds can be eliminated, the limit of detection can be reduced to 0·5 mg/litre by injecting 200 μl of sample and preliminary removal of carbon dioxide.

The complete combustion of the injected samples is usually achieved without difficulty, particularly if pure oxygen is used as the carrier gas instead of air. In the case of certain chemical waste waters the recommended catalyst, cobalt oxide on asbestos, cannot be used and must be replaced by another oxidation catalyst which is resistant to acids and salts. Based on our experience, the following catalysts which have proved themselves in micro-analysis, can be recommended; they are packed in the combustion tube as shown in Fig. 4:

(a) silver tungstate on magnesia, Perkin–Elmer Cat. No. 240-1344;
(b) silver oxide and silver tungstate on Chromosorb P 30/60 mesh, Perkin–Elmer Cat. No. 240-0113;
(c) manganese(IV) oxide on granulated support, Merck Cat. No. 5953;
(d) copper(II) oxide, granulated, for elemental analysis, Merck Cat. No. 2764.

The most effective catalyst must be chosen for each particular case. It is

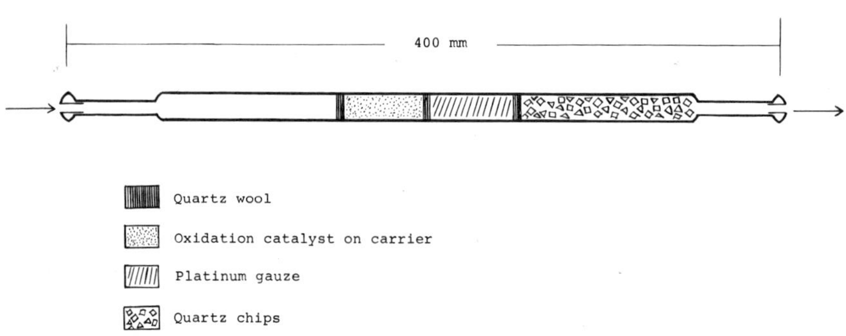

FIG. 4. Combustion tube filling.

therefore advantageous if solutions containing the substances or substance classes present in the water are available for testing purposes.

A high acid and salt content of the water affects not only the efficiency and working life of the tube packing, but also leads to an overall contamination of the equipment because of salt deposition. Large amounts of nitrogen oxides are evolved in the case of waste waters containing large amounts of nitric acid or nitrates; this leads to high results, because of the weak absorption of these oxides at the absorption wavelength of carbon dioxide. If their carbon content permits it, such water samples which affect the whole equipment should be diluted prior to analysis.

The time required for a single determination is 2–4 minutes, depending on the amount of sample injected. A total time of about 100 minutes is required for 10 duplicate TOC determinations including calibration. Thus, the equipment can be used advantageously in cases where many waste water samples have to be analyzed daily.

B. W. C. Heraeus, Merz Rapid C "TOC analyzer"

In contrast to the combustion procedures which allow only the injection of small water samples of 20–200 μl, the TOC analyzer developed by Merz[20] permits the combustion of up to 5-ml samples. This is achieved by the use of a vertical combustion tube equipped with a sample injection device, designed for the slow injection of measured sample volumes, and by the continuous automatic titration of the carbon dioxide, absorbed in a nonaqueous solution, with tetrabutylammonium hydroxide (TBAH). The calibration plot is linear so that an appropriate combination of sample volume and normality of the TBAH solution makes it possible to cover a wide range of cargo concentrations from 0·5 to 10000 mg C/litre and to select the most suitable approach for the analysis of widely varying water samples.

The TOC content is determined by the difference method, whereby the TC value is determined in the combustion module and the IC value in the degassing unit of the equipment.

The design and operation of the combustion module are shown in Fig. 5.

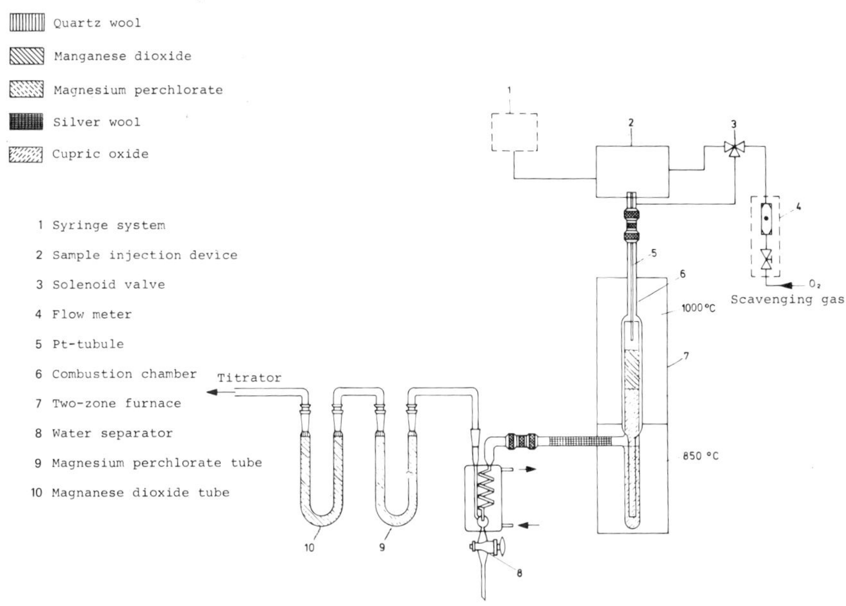

FIG. 5. Heraeus, TOC-Analyzer, according to Merz Rapid C. Flow diagram of combustion unit.

The sample (0·5–5 ml) is aspirated into a syringe and injected slowly during 1–2 min through a narrow platinum capillary into the combustion space, with the injection stopcock open. The injection stopcock is then closed and the liquid remaining in the platinum capillary flushed out by short diversion of the purified carbon dioxide-free oxygen flow through the injection device into the combustion tube. The lower part of the combustion tube, which is heated to 850°C, is packed with copper oxide as the combustion catalyst. The upper part, which is heated to 1000°C, represents the combustion chamber and is packed with manganese dioxide for the absorption of salt residues. After injection of the sample, the water vapour which emerges from the tube passes over a silver gauze placed at the furnace outlet which removes halogen and sulphur dioxide and is condensed in a cooling coil. The carbon dioxide formed in the combustion is flushed by the carrier gas through a drying tube and a tube for the absorption of nitrogen oxides, and finally reaches the titration cell where it is absorbed in a solution of ethanolamine in dimethylformamide and continuously titrated with 0·02 M TBAH solution with thymolphthalein as the indicator. The principles of automatic photometric titration have been described previously.[12, 13]

To determine the inorganic carbon, 0·5–5 ml of sample are injected into a degassing vessel which contains a solution of hydrogen peroxide, silver sulphate, and dilute sulphuric acid. The carrier gas transports the carbon dioxide liberated from carbonates through a drying tube, a cold trap and a nitrogen oxide absorber packed with manganese dioxide to the photometric titrator, where it is titrated, as after combustion, with 0·02 M TBAH solution. Interfering acids such as hydrochloric and sulphurous acids, hydrogen sulphide, or halogens are precipitated by the silver sulphate present in the degassing solution; volatile acids are condensed in the cold trap.

The determination of TC by combustion and of IC by degassing can be carried out in sequence or, if two separate photometric titrators are used, simultaneously. In the latter case, the duration of a TOC determination with automatic timing is 8 minutes. If the inorganic carbon is removed from the sample before the combustion by acidification and purging, the TOC value is obtained directly from the combustion. The analyzer can be further automated by incorporating an automatic sampler and data handling equipment, such as a calculator, printer, tape puncher, etc.

C. H. Maihak, "TOC–UNOR"

The TOC–UNOR developed by Axt[21] is particularly suitable for carbon determination in surface and drinking waters with a low carbon content, because of continuous sample supply and recirculation of the carrier gas, as well as the high sensitivity of the IR detector with working ranges from 0–1 mg/litre to 0–10 mg/litre. The analyzer can also be used on waste waters with high carbon content if it is adapted for the higher concentration ranges required in this case. An advantage of the principle on which the analyzer is based for the TOC determination in industrial waste waters with a high level of volatile organic constituents is that these volatile substances are not lost in the removal of the inorganic carbon, but are returned to the combustion chamber and the organic carbon, thus avoiding a separate determination or a TOC determination by the difference method.

The flow diagram in Fig. 6 shows the design and operation of the analyzer.

The acidified water sample is supplied to the analyzer continuously through the dosage capillary at a rate of 20–80 ml/h. The carbon dioxide liberated by the acid is removed with an efficiency of 99·95% by the counter-current flow of air as the water flows down the inner wall of the degassing tube. The water, from which the inorganic carbon has been removed, drops

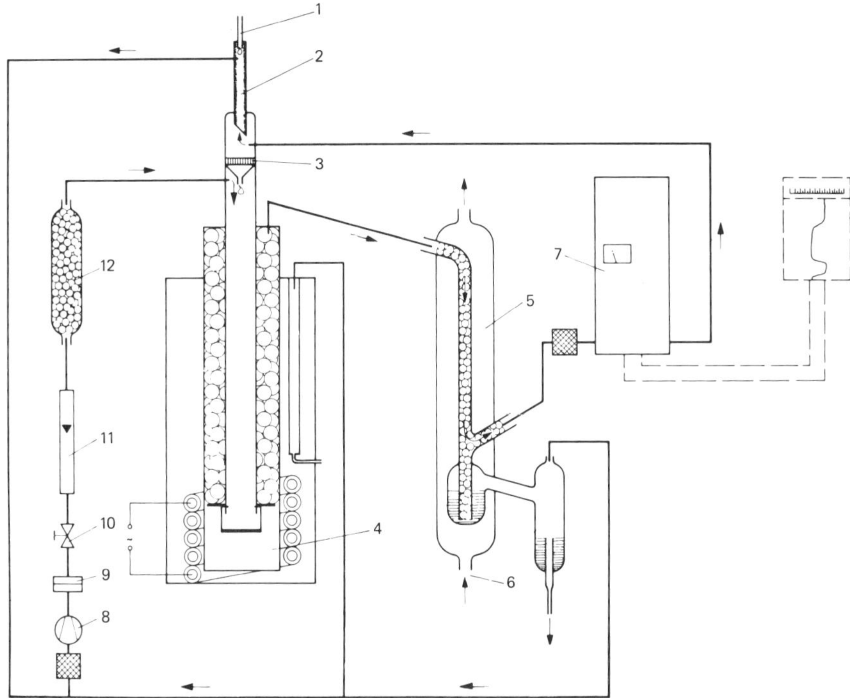

FIG. 6. Maihak, TOC-UNOR; Flow diagram. 1) Sample inlet; 2) Degassing tube; 3) Porous frit; 4) Combustion chamber; 5) Steam condensor; 6) Cooling water inlet; 7) IR-Analyzer; 8) Membrane pump; 9) Flow controller; 10) Valve; 11) Flowmeter; 12) Carbon dioxide absorber.

through a glass frit into the combustion chamber where the organic substances are vaporized and combusted on the oxidation catalyst. The steam leaving the combustion chamber together with the air supplied for the combustion, is condensed by means of a condenser and leaves the analyzer through the condensate outlet. The carbon dioxide formed in the combustion is swept by the carrier gas into the IR analyzer, giving a continuous trace of the carbon dioxide concentration which after an initial period reaches a constant level. The air stream leaving the IR analyzer is returned to the degassing tube and serves to purge the acidified water entering the analyzer. The volatile organic substances which are purged together with the carbon dioxide, are aspirated by the membrane pump and returned to the combustion chamber via the carbon dioxide absorber.

D. Carlo Erba Model 420 total carbon monitor

This TOC monitor is designed primarily for the fully automatic monitoring of waters and water treatment plants, and is also suitable for the analysis of discrete samples. It is based on the following principle (Fig. 7). The sample is drawn through continuously by means of a peristaltic pump and is brought into contact with 2 M nitric acid, transported by the second pump channel.

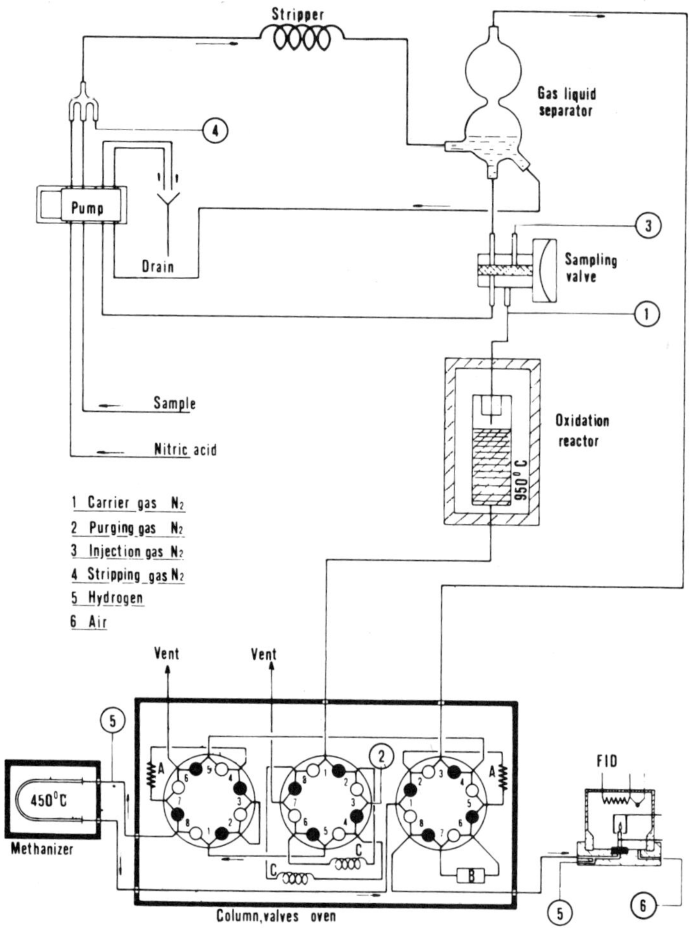

FIG. 7. Carlo Erba, Total Carbon Monitor Mod. 420; Flow diagram.

A stream of nitrogen carries the acidified sample through a glass coil where the liquid is mixed thoroughly with the carrier gas. The carbon dioxide liberated from the inorganic carbon, as well as volatile organic substances, pass into the gaseous phase which is then separated from the liquid phase in the adjacent gas–liquid separator. A portion of the water that has been freed from carbon dioxide and volatile organic carbon is aspirated through the slide valve of the sample injection device, and the rest is discarded as overflow. By shifting the slide valve, the bore of the valve, which has a volume of 20 μl, is pushed into a stream of nitrogen which then carries the measured sample volume into the oxidation reactor, heated to 950°C. The carbon dioxide formed in the combustion on copper oxide leaves the combustion furnace, is separated from the water in a column, mixed with a sufficient volume of hydrogen and converted to methane by hydrogenation over a nickel catalyst at 450°C. The gas emerging from the CO_2-to-methane converter produces a signal in the flame ionization detector (FID) which is proportional to the carbon concentration of the combustion gas.

The carrier gas, separated in the gas–liquid separator, containing the carbon dioxide and volatile organic substances passes through two 1·5-ml sampling loops arranged in sequence and is vented into the atmosphere. The two sampling loops are included separately, one after the other, into a nitrogen stream and flushed out by means of automatically controlled multiport valves. The contents of the first sample loop are supplied directly to the FID and a signal proportional to the amount of volatile organic carbon present in the water sample is obtained. The contents of the second sampling loop pass through the carbon dioxide-to-methane converter before they reach the FID, which then records a signal which is proportional to the sum of inorganic and volatile organic carbon. Hence, the automatic analysis cycle with a duration of 5 minutes produces three signal peaks which are recorded directly or as peak area integrals, indicated by a corresponding recorder pen deflection. After calibration of the three signals with suitable substances, the analyzer gives the following values:

1. dissolved, nonvolatile organic carbon;
2. volatile organic carbon;
3. sum of inorganic and volatile organic carbon.

The TOC content corresponds to the sum of the first and second value, and the IC content to the difference between the third and second values.

The analyzer can be equipped with a special data-handling module, which records and prints out directly the TOC, TC and IC values.

E. Dohrmann DC-52 TOC analyzer

The instrument can be applied to the determination of organic carbon in surface, sea, and drinking water, in industrial waste waters with high and low carbon contents, in sludges and even in sediments. The versatility of the analyzer is achieved by:

1. pyrolysis and combustion of the samples in a platinum boat, thus reducing contamination of the pyrolysis tube by salts;
2. the removal of inorganic carbon without loss of volatile organic carbon;
3. conversion of carbon dioxide and organic pyrolysis products to methane and its determination with a sensitive flame ionization detector (FID) whose response is linear over a wide concentration range;[16]
4. the provision of two different operating modes which can be chosen according to the carbon content of the sample.

The "TOC-Direct" mode is recommended for waste waters with carbon contents higher than 10 mg/litre and in the presence of volatile organic substances. The "TOC-Sparge" mode, which offers a detection limit of 0·05 mg C/litre, is suitable for surface waters with carbon contents of less than 10 mg/litre.

The "TOC-Direct" Mode: 10–30 µl of the acidified water sample are injected with a syringe through a side-arm on the pyrolysis tube into a platinum boat which contains a cobalt oxide catalyst to assist the combustion. An automatic transport device pushes the platinum boat in the pyrolysis tube into the vaporization zone heated to 110°C where the water vapour and the carbon dioxide liberated from the carbonates are picked up by the helium carrier gas, passed through a Porapak Q column heated to 60°C and vented to the atmosphere (Fig. 8(a)). When the vaporization step of the analysis cycle is completed, the volatile organic substances retained on the column are stripped from the column by heating to 130°C and flushed by helium flowing in the opposite direction back into the pyrolysis tube (Fig. 8(b)). There they are mixed with hydrogen gas and converted to methane at 350°C in a reaction zone packed with nickel catalyst. The methane in the helium carrier gas produces a linear FID signal, the peak area of which is integrated and displayed digitally. The value obtained corresponds to the volatile carbon content of the sample.

In the following pyrolysis step of the analysis cycle, the sample boat is pushed into a pyrolysis zone heated to 850°C where the substance remaining in the boat is evaporated or combusted. The nickel catalyst flushed with hydrogen reduces the pyrolysis gases and the carbon dioxide completely to methane. The FID signal which is proportional to the methane concentration is integrated, and the value corresponding to the nonvolatile organic carbon content is added to the volatile carbon content. The digital display of the instrument represents the TOC content of the sample.

The "TOC-Sparge" mode. The "TOC-Direct" Mode is not suited to the determination of TOC in surface waters. The TOC contents are usually lower than 5 mg/litre and it is difficult to achieve the necessary complete

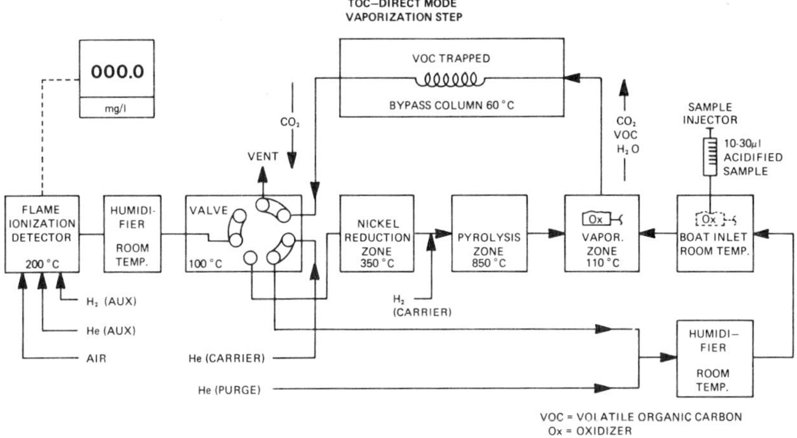

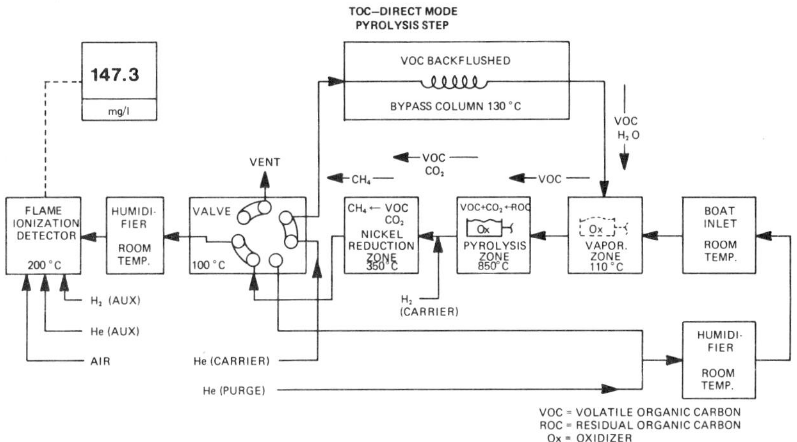

FIG. 8. Dohrmann, TOC-Analyzer DC-52; Flow diagram of TOC-direct mode. (a) Vaporization step. (b) Pyrolysis step.

removal of the inorganic carbon which is usually present in large amounts in relation to the organic carbon. The "TOC-Sparge" mode is therefore based on the removal of the inorganic carbon by stripping of the sample acidified to pH 2 with a stream of nitrogen or helium. The CO_2-free sample (100 μl) is injected through an injection opening without a septum into the platinum boat containing the cobalt oxide catalyst which is then pushed into the vaporization zone heated to 200°C. The substances which are volatilized at this temperature are converted, after removal of the water vapour, over a nickel catalyst to methane, which is again supplied to the FID. After appearance of the signal proportional to the carbon content, the boat is pushed further into the pyrolysis zone heated to 850°C, where the organic residue remaining in the boat is vaporized or combusted to carbon dioxide. The pyrolysis and combustion gases are again converted to methane. The value obtained by integration of the FID signal represents, together with the value for the volatile constituents, the TOC content of the sample in mg/litre.

F. Oceanography International Model 0524 total carbon system

The analyzer provides two different procedures for the determination of carbon. In the first procedure, which is intended for carbon contents from 10–50 000 mg/litre and which corresponds to the method of Van Hall and co-workers,[11] a sample volume of 20 to 100 μl, depending on the carbon content of the sample, is injected into a combustion tube and combusted at 950°C in a stream of oxygen. The combustion gases are swept into a non-dispersive IR analyzer and its output signal is recorded on a recorder or integrated by the built-in integrator. The digitally displayed peak area integral corresponds to the total carbon content of the sample injected. For the TOC determination, the inorganic carbon must be removed from the sample by acidification and stripping of the carbon dioxide formed.

In the second procedure, samples of up to 10 ml are subjected to wet-chemical digestion in glass ampoules, thus making the method particularly suitable for the analysis of waters with carbon contents in the range 0·1–10 mg/litre. In this ampoule procedure, which is also advantageous in the case of acid or salt-containing waters, an analysis is carried out as follows: 10 ml of sample are placed in a 10-ml glass ampoule and 0·2 g of potassium peroxydisulphate and 0·25 ml of 6% phosphoric acid added. A special ampoule preparation and sealing device is used to remove the inorganic carbon from the ampoule by purging with oxygen and to seal it with a microburner incorporated in the device. A large number of ampoules prepared in this way are kept in a pressure vessel at 175°C for eight hours. After completed oxidation and cooling, an ampoule is attached to the analyzer and the seal broken so that a carrier gas stream can flush the carbon dioxide formed from the ampoule and transport it to the IR analyzer. After subtraction of a reagent blank from the peak integral, the TOC content of the sample is obtained from a calibration plot, established beforehand. In a variant of the ampoule method described by Baldwin and McAtee,[7] the oxidation rate is increased significantly by the addition of silver nitrate to the ampoule contents, so that the oxidation, at least in the case of standard solutions and pure waters, can be completed in 30 minutes, even at room temperature. The disadvantage of both variants is that CO_2-free filling, purging, and sealing of the ampoules requires very careful handling and that the time required for a determination, even with the fast variant, is of the order of 80 minutes.

G. H. Wösthoff "Hydromat-TOC"

The analyzer is based on the method developed by Schierjott, Bleier, and Malissa[22] for the simultaneous continuous determination of the chemical oxygen demand and the organic carbon content. The method employs wet-chemical oxidation of the organic constituents with a mixture of dichromate and sulphuric acid. It is therefore not affected by high salt or acid contents and is particularly suited for the analysis of industrial waste waters. The method can be used for the analysis of discrete samples as well as for the continuous monitoring of waste waters.

An analysis is carried out as follows: the sample drawn by means of a peristaltic pump is continuously mixed with concentrated sulphuric acid in the ratio 1:1 and contacted with a strong countercurrent air stream, which continuously removes the dissolved carbon dioxide and that liberated from carbonates in the sample stream. A portion of the CO_2-free sample is picked up by another channel of the peristaltic pump, and mixed in the ratio 1:1 with the oxidant mixture consisting of concentrated sulphuric acid, potassium dichromate and a silver sulphate catalyst. After segmentation with CO_2-free air, the mixture is pumped through a glass coil heated to 165°C. The residence time in the glass coil is 15 minutes, so that a complete digestion of the organic substances is achieved. After completed oxidation and separation of the oxidation mixture in a gas separator, an aliquot portion of the carbon dioxide formed is picked up by a channel of the peristaltic pump and flushed with a carrier gas into a conductivity cell, where the carbon dioxide present in the gas stream is absorbed in sodium hydroxide solution. The sodium hydroxide solution is continuously pumped through the measuring cell and its conductivity is measured in the reference channel before absorption of the carbon dioxide and in the sample channel after absorption of carbon dioxide. The change in conductivity due to the absorption of carbon dioxide is a direct measure of the TOC content of the sample. Chlorine interferes in the determination and is removed from the carrier gas stream partly by purging before the oxidation, and partly before the measurement by means of a chlorine absorber.

H. H. Maihak UV–DOC–UNOR

The UV–DOC analyzer developed by Wölfel[9] is based on the oxidation of aqueous solutions by oxygen under the influence of UV light and measurement of the carbon dioxide formed in a nondispersive IR analyzer. The large reactor volume makes it possible to vary the sample size within wide limits from 50–20 000 μl; this provides a low detection limit of 0·001 mg C/litre and a good reproducibility in the measurement ranges from 0·1 mg C/litre to 10 g C/litre. The low reagent consumption avoids the need for blank values and the equipment is not contaminated by oxidation residues. The analyzer is simple to operate and quickly ready for use, and can be operated manually as well as automatically, with and without a sample changer. All these characteristics make the analyzer particularly suitable for the analysis of surface and general-purpose waters.

The central item of the analyzer, the oxidation vessel, consists of a cylindrical Pyrex reaction vessel of 300 ml capacity, into which a low-pressure mercury lamp made of quartz is inserted. The reaction vessel which is thermostated at 40°C contains about 300 ml of carbon-free water, acidified with phosphoric acid to pH 2. A sample is introduced into this solution and the dissolved organic substances are oxidized with a finely dispersed stream of oxygen bubbling through the solution at a flow rate of 0·5–1 litre/min in the presence of UV radiation from the mercury lamp. The gas stream leaving the reaction vessel passes, after removal of the entrained water in a condenser, through a tubular furnace kept at 600°C where the entrained volatile organic substances and oxidation products are combusted and ozone formed by the irradiation is decomposed.

The carbon dioxide concentration in the oxygen carrier gas is determined with a nondispersive IR analyzer. The signal in peak form is recorded and integrated. After calibration with a standard solution, the integrator displays the carbon content of the sample in digital form.

Before the start of a series of analyses, the reaction solution is treated with oxygen under UV light until the carbon present in the solution is removed and the recorder of the IR analyzer shows a steady baseline. With the UV lamp switched off, the water sample to be analysed is injected into the CO_2-free solution and the inorganic carbon purged for 2 minutes and recorded. The UV lamp is switched on, the oxidation starts, and its course can be followed by the recorder pen deflection. Depending on the type of sample, the oxidation takes 3 to 12 minutes after which time the recorder pen returns to the baseline and the carbon content is displayed after a preselected integration period.

Volatile organic constituents are combusted during purging of the inorganic carbon in the afterburner and determined together with the inorganic carbon. A repeated purging, with by-passing of the afterburner, gives the inorganic carbon alone, so that the volatile organic carbon can be determined from the difference between the two determinations.

The method has the disadvantage that undissolved organic carbon cannot be determined and that certain substances are difficult to oxidize, so that in this case the integration period must be extended. The interference in the oxidation caused by chloride ions can be eliminated, according to Wölfel and Sontheimer,[9] by the addition of a sufficient amount of mercury(II) sulphate.

The equipment described is manufactured by the following firms:

Beckman Instruments, Inc.
Process Instruments Division, Fullerton, California 92634

Ionics Inc.
65 Grove Street, Watertown, Massachusetts 02172

W. C. Heraeus GmbH.
Werksgruppe Elektrowärme, D-6450 Hanau, P.O. Box 169

H. Maihak AG.
D-2 Hamburg 60, Semperstrasse 38

Carlo Erba Strumentazione
Divisione Apparecchi Scientifici, I-20090, Rodano-Milano, P.O. Box 4342, 20100, Milano

Envirotech/Dohrmann Corp.
3240 Scott Boulevard, Santa Clara, California 95050

Oceanography International Corporation
College Station, Texas

H. Wösthoff oHG.
D-463 Bochum, Hagenstrasse 30

We wish to express our gratitude particularly to the following firms for their assistance and the supply of material for illustrations:

Beckman Instruments, Carlo Erba, Envirotech/Dohrmann, W. C. Heraeus and H. Maihak.

IV. REFERENCES

1. K. Grob, *Journal of Chromatography*, **84** (1973), 255–273.
2. R. A. Dobbs, R. H. Wise and R. B. Dean, *Water Res.* **6** (1972), 1173–1180.
3. E. Abrahamczik, G. Groh, W. Huber and Fr. Kraus, *Vom Wasser*, **37** (1970), 82–91.
4. D. D. Van Slyke, *Anal. Chem.* **26** (1954), 1706–1712.
5. F. Ehrenberger, *Z.f. Wasser- und Abwasser-Forschung,* **8** (1975), 75–81.
6. J. Katz, S. Abraham and N. Baker *Anal. Chem.* **26** (1954), 1503–1504.

7. J. M. Baldwin and R. E. McAtee, *Microchem. J.* **19** (1974), 179–190.
8. M. Erhardt, *Deep Sea Res.* **16** (1969), 393.
9. P. Wölfel and H. Sontheimer, *Vom Wasser* **43** (1974), 315–325.
10. P. D. Goulden and P. Brooksbank, *Anal. Chem.* **47** (1975), 1943–1946.
11. C. E. Van Hall, J. Safranko and V. A. Stenger, *Anal. Chem.* **35** (1963), 315–319.
12. O. I. Snoek and P. Gouverneur, *Anal. Chim. Acta,* **39** (1967), 463.
13. W. Merz, *Anal. Chim. Acta,* **48** (1969), 381.
14. W. Schmidts and W. Bartscher, *Z. Anal. Chem.* **181** (1961), 54–59.
15. C. E. Van Hall and V. A. Stenger, *Water and Sewage Works,* **111** (1964), 266.
16. R. A. Dobbs, R. H. Wise and R. B. Dean *Anal. Chem.* **39** (1967), 1255–1258.
17. F. Ehrenberger, *Z. Anal. Chem.* **267** (1973), 17–21.
18. J. Ruzicka, U. Fiedler and E. H. Hansen, *Anal. Chim. Acta,* **74** (1975), 423–435.
19. C. E. Van Hall and V. A. Stenger, *Anal. Chem.* **39** (1967), 503–507.
20. W. Merz, *G-I-T, Fachz. Lab.* **19** (1975), 293–301.
21. G. Axt, *Vom Wasser,* **36**, 1969 (1970), 328–339.
22. G. Schierjott, H. Bleier and H. Malissa *Vom Wasser* **39** (1972), 1–7.

5. THE DETERMINATION OF NITROGEN BY THE MERZ MODIFICATION OF THE DUMAS METHOD

P. I. BREWER
Esso Research Centre,
Abingdon, Oxfordshire.

I. INTRODUCTION

In the classical Dumas method for determining nitrogen the sample is mixed with copper(II) oxide and heated strongly. The resulting pyrolysis products are further oxidized by passing over heated copper(II) oxide in a stream of carbon dioxide. Any oxygen present is then removed by passing over heated copper which also serves to reduce oxides of nitrogen. The resulting gas which should contain only carbon dioxide and nitrogen passes to a nitrometer where the carbon dioxide is absorbed by potassium hydroxide and the nitrogen is measured. On the micro-scale a determination takes about 1 hour. A great deal of experience and skill are needed to obtain satisfactory results; the main difficulty is in oxidizing the sample completely. Although automating the apparatus and using catalysts or other oxygen donors improves the precision, it does not completely eliminate these difficulties. Attempts have been made by several investigators to burn the sample in a limited amount of oxygen and then sweep the combustion products through heated copper(II) oxide and copper as before.[1]

An apparatus incorporating this modification was described by Merz[2] in 1968. The apparatus was partly automated, the supply of oxygen and carbon dioxide to the combustion tube being controlled by means of magnetic valves. Once the cycle had been started the operation was automatic except for adding the sample. A single determination could be made in about 4 minutes which was considerably more rapid than the classical Dumas procedure. The apparatus was later modified for special purposes: a model was developed for small quantities of nitrogen and another modification was intended for trace quantities of nitrogen in aqueous solutions. A fully automatic instrument was also developed. These instruments are now made by W. C. Heraeus of Hanau and are marketed under the name of the "Micro-Rapid-N".[3] A photograph of the standard apparatus is shown in Fig. 1.

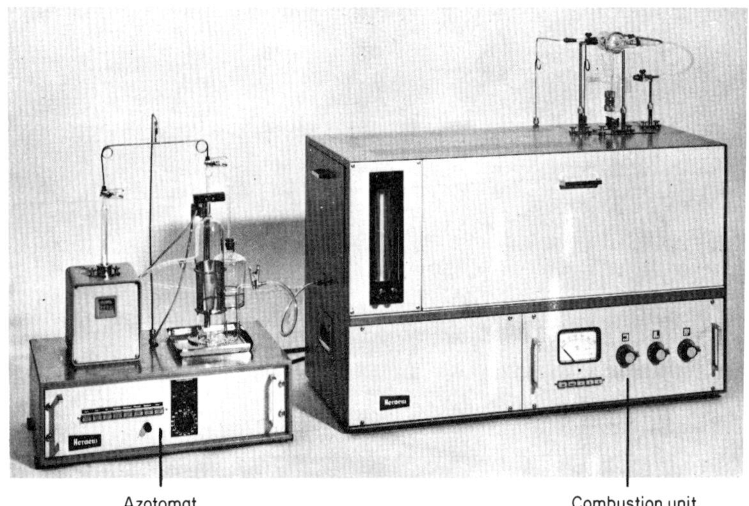

FIG. 1. The standard Micro-Rapid-N apparatus.

II. APPARATUS

A. The combustion train

A diagrammatic view of the combustion train is shown in Fig. 2. The sample admission cock A is a large three-way glass stopcock located on top of the silica tube B. The sample is weighed into a suitable container which

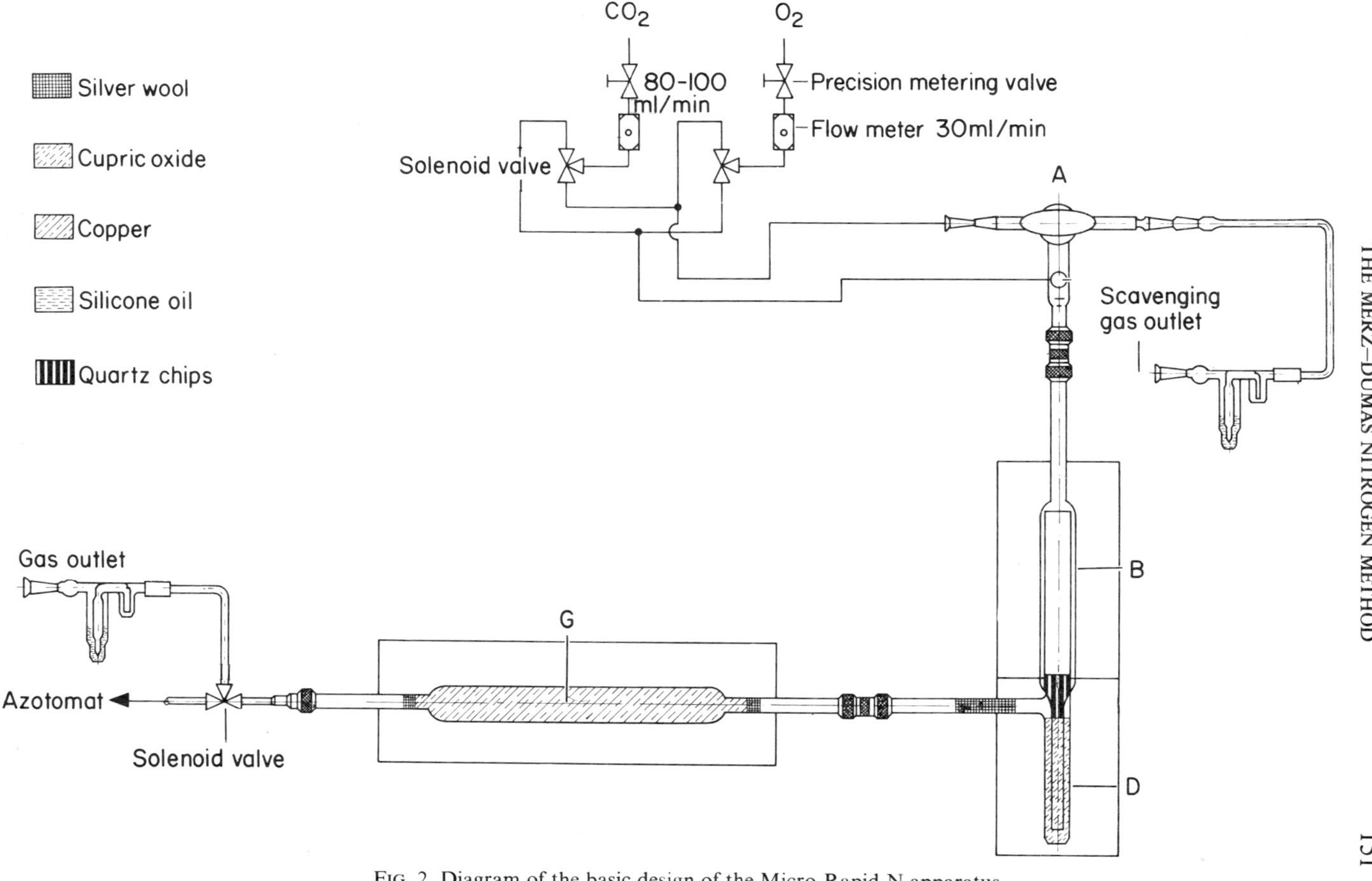

FIG. 2. Diagram of the basic design of the Micro-Rapid-N apparatus.

is then placed in the hole in the plug. After removing the air in the plug by purging with oxygen or carbon dioxide, the sample is dropped into the tube by turning the plug through 90°. The plug is then returned to its original position. The upper part of tube B is empty except for a silica sleeve which serves to protect the tube from inorganic material introduced with the sample. The lower part of the tube is filled with copper(II) oxide and is double-walled to provide a longer path length for the pyrolysis products. A layer of silica chips rests on the top of the copper(II) oxide. The upper part of the tube B is maintained at 1000–1050°C and the lower part D at 850–900°C. The side-arm of the tube is packed with silver wool to remove halogens. The horizontal silica tube G is filled with copper and is heated to 500–550°C. The packing in this tube is also held in position with silver wool. Connections between the reaction tubes and between the tube B and the stop-cock are made by metal connectors with rubber O-rings.

Gases leaving the reduction tube pass either to the atmosphere via a bubbler or to the nitrometer ("Azotomat") which is shown in Fig. 3. The absorption chamber of the nitrometer is filled with 50% potassium hydroxide solution. At the bottom of the chamber is a magnetic stirrer which disperses the incoming gas. The top of the chamber is connected to a motorized digital burette by means of narrow-bore metal tubing and connections are made to the burette and absorption chamber with spherical joints with O-rings.

The burette motor is controlled by a photoelectric level-sensing device attached to the capillary on top of the absorption chamber. This capillary terminates in a bulb which contains a metal electrode. If the photoelectric sensor fails and liquid flows through the capillary when it touches the electrode, a valve connecting the system to the atmosphere is opened and a buzzer sounds.

B. Sample preparation

Non-volatile solids and viscous liquids are weighed into aluminium boats and copper(II) oxide added. This helps in the oxidation and soaks up liquid samples so that they are not squeezed out of the boat when it is folded. Lunder[4] found that copper(I) oxide was a more efficient combustion aid in the determination of nitrogen in some foodstuffs. Volatile liquids are weighed into glass capillary tubes.

C. Automated sequence of operation

After the furnaces have reached their operating temperatures, and the oxygen and carbon dioxide flow-rates have been adjusted, the sample is placed in the admission cock. The start button is then pressed; after this, the

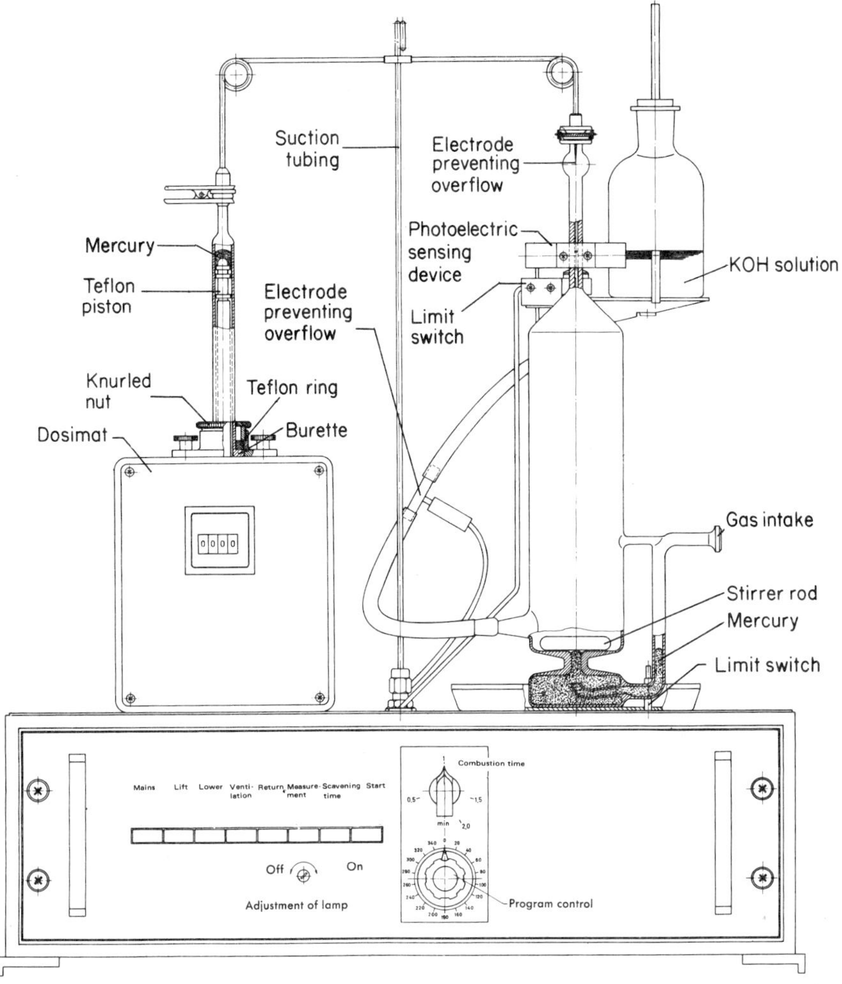

FIG. 3. The Azotomat.

programme proceeds automatically with the exception of placing the sample in the furnace. The programme is divided into 360 divisions, the sequence is set out below.

0° (a) The nitrometer levelling bottle is lowered and the magnetic stirrer started.

(b) The gas scavenging the admission cock changes from oxygen carbon dioxide.

(c) The gas passing to the combustion tube changes from carbon dioxide to oxygen. The duration of the oxygen flow-time, i.e. the combustion period can be varied from 30 to 120 seconds by means of a control on the front of the instrument.

(d) The gas from the reduction furnace is routed to the nitrometer.

(e) After approximately half the combustion period has elapsed, the sample is dropped into the furnace by turning the admission cock through 90° and then back again to its original position. The next sample is then placed in the admission cock so that it is being purged while the combustion of the first sample takes place.

40° (a) The gas passing through the admission cock changes from CO_2 to O_2.

(b) The gas passing to the combustion tube changes from oxygen to carbon dioxide.

260° (a) The gas from the reduction furnace is switched from the nitrometer to the atmosphere via the bubbler.

(b) The nitrometer levelling bottle is raised and the magnetic stirrer switched off.

290° The nitrogen is passed into the burette.

300° About 0·25 ml of nitrogen is returned to the absorption vessel.

320° The nitrogen is again passed into the burette. The burette reading gives the final volume of nitrogen.

360° The button marked "Ventilation" is depressed manually. This expels the nitrogen from the burette and the burette piston and potassium hydroxide are returned to their original positions. When the button is disconnected the apparatus is ready for the next run.

It will be noted that after a preliminary measurement of the carbon dioxide, about 0·25 ml is passed back into the potassium hydroxide solution and then back again into the burette. This enables the efficiency of the carbon dioxide absorption to be checked and also corrects for the presence of bubbles that may have been left in the potassium hydroxide. For this reason, it is advisable to add a known small amount of air before making a blank determination or when the amount of nitrogen present is small.

D. Special modifications

The apparatus shown in Fig. 1 is used for compounds containing more than 1% nitrogen with a sample of up to about 10 mg and a combustion period of 1 minute. For low nitrogen contents, the apparatus is modified to cope with larger samples. An extra reaction tube is placed between the vertical tube D and the reduction tube G. The silica tube is half-filled with copper (II) oxide held in position with silica wool. The other half is left empty to provide a buffer volume. This reaction tube is placed in a separate enclosure, which is shown labelled H in Fig. 4. The maximum amount of sample that can be handled with this modification varies between 25 and 100 mg depending on the combustion characteristics of the sample. With this amount of sample the combustion period would be set to the maximum of 2 minutes.

For the determination of nitrogen in aqueous solutions, the extra reduction furnace is incorporated and the admission cock is replaced by a special entrance block which allows the injection of up to 1 ml of solution with a syringe (Fig. 5). A water-cooled condenser is placed after the copper(II) oxide tube and a mercury trap between the reduction tube and the nitrometer. With these modifications the lower limit is about 0·005% nitrogen.

It is possible to obtain a fully automated version of the Micro-Rapid N apparatus; up to 50 weighed samples can be loaded into the apparatus which is then left to run without supervision.

III. REAGENTS

The purity of the carbon dioxide and oxygen is important, since the greater the nitrogen and argon content of these gases, the greater the blank. High blanks lead to poorer repeatability and a decrease in sensitivity. Carbon dioxide with a purity of >99·9% is available from several sources, and 99·9% oxygen can be obtained from the Special Gases Division of the British Oxygen Company.

The combustion tubes D and H are filled with wire-form copper(II) oxide 3–5 mm in length and about 0·6 mm diameter. The copper in tube G is also in the wire form of about the same size as the copper oxide. The extent of the oxidation of the copper in the reduction tube can be followed by noting the change in colour. It often happens that oxidation proceeds more rapidly along the top of the packing; in this case the packing should be made more even by occasionally rotating the tube through 180°. When about two-thirds of the copper has been oxidized it should be reduced by passing a stream of nitrogen saturated with methanol through the heated furnace. Both the silica combustion tubes filled with copper(II) oxide and copper decrease in efficiency with time due to channelling; when this happens they should be repacked or replaced if devitrified. It is usually impossible to remove the old packings mechanically and so they must be dissolved in a suitable reagent such as aqua regia. In the case of tubes G and H, this is probably not worth the labour involved.

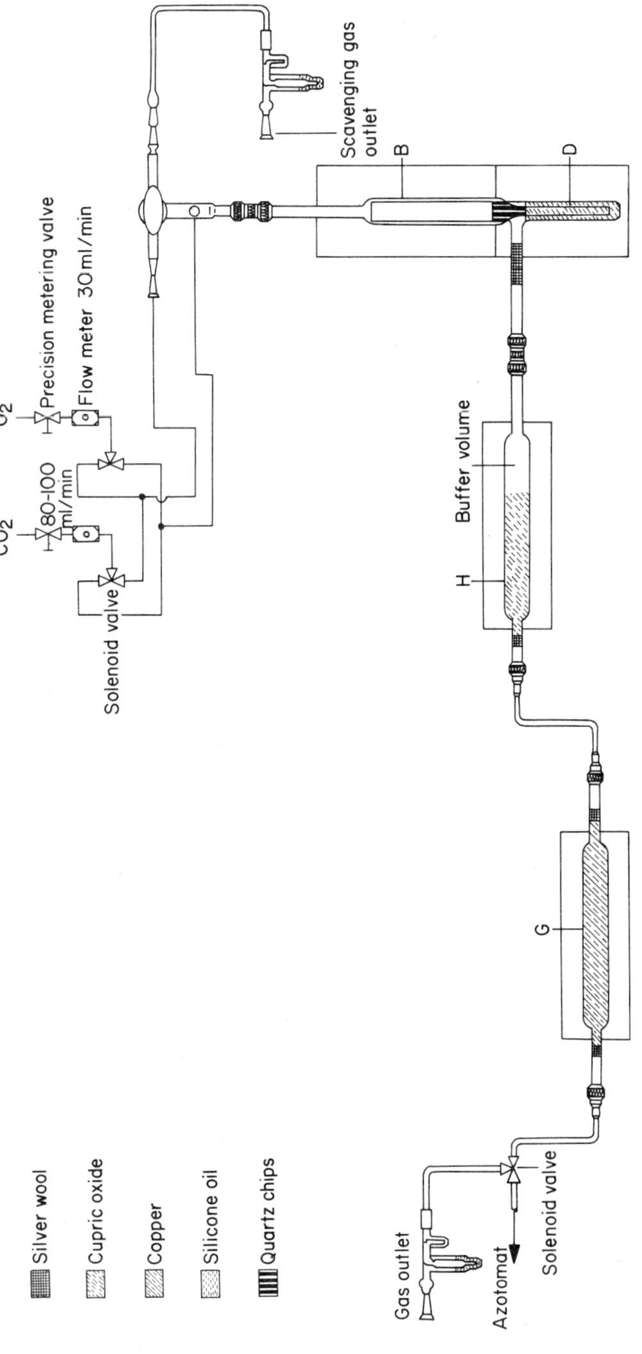

Fig. 4. Diagram of the modified Micro-Rapid-N apparatus for trace determinations.

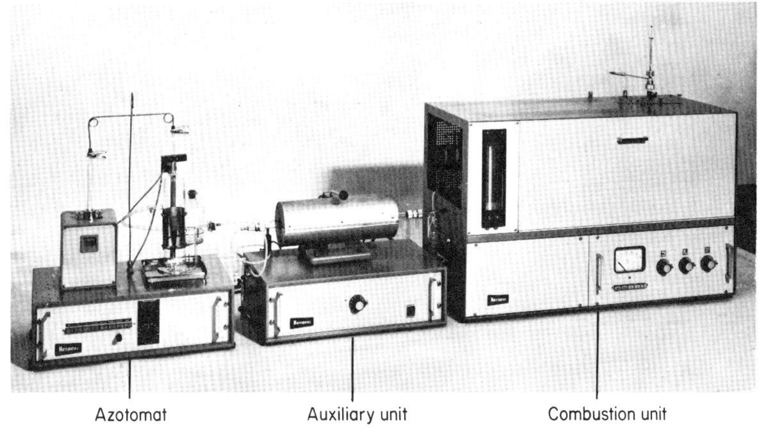

FIG. 5. The apparatus modified for aqueous solutions.

The potassium hydroxide solution (50% w/v) in the nitrometer gradually becomes exhausted and must be renewed about every 100 determinations. The freshly prepared solution should be either filtered or allowed to stand, as particulate matter interferes with the operation of the detector.

IV. PROCEDURE

It is advisable to check the apparatus for leaks before each day's run. The carbon dioxide flow-rate is set at about 80 ml/minute, and the gas outlet at the bubbler attached to the reduction tube is closed by means of a rubber bung. The time taken for the flow rate to be reduced to one-third of its original value is then noted. If this is less than five minutes, the apparatus is sufficiently leak-free for use. If a leak is indicated, it should be located by applying soap solution to the joints with a fine brush.

The furnace temperature controls are set as follows: furnace B, 1000°C; furnace D, 920°C; furnace H, 860°C; furnace G, 560°C. The furnaces are energized and allowed to heat to the set temperature; this will take about $2\frac{1}{2}$ hours. If possible the furnaces should be left on permanently; this tends to give a smaller and more reproducible blank. The oxygen flow-rate is set to 40 ml/min and the combustion period to the required value. The nitrometer is energized and the zero adjusted. This is done by taking a small amount of air into the nitrometer by lowering the levelling bottle and momentarily depressing the "Ventilation" button. The volume of the air is then measured, returned to the absorption vessel and again measured; these values should agree. On pressing the "Ventilation" button, there should be no movement of

F*

the meniscus in the capillary tube. If movement does take place, the position of the photocell must be adjusted.

An initial run should be made without adding a sample by pressing the "Start" button; the final reading is ignored. A blank determination is then made by weighing about 20 mg of benzoic acid into an aluminium boat and adding about 0·3 g of copper(II) oxide. The sides of the boat are then folded over and compressed by the fingers so that the minimum of free space remains (gloves should be worn for this operation). A small quantity of air should be taken into the nitrometer (about 0·3 ml is sufficient) and its volume measured. The boat is placed in the admission cock and after 2 minutes the cycle started. After half the combustion period has passed, the boat is placed in the furnace by turning the admission cock through 90° and then back to its original position. At the end of the cycle the final burette reading is noted. The volume of the air originally present is subtracted from this to obtain the blank. The blank should be less than 0·05 ml.

In an actual determination the weight of sample taken will be governed by the nitrogen content of the sample. The following guide can be used:

for 10–40% N, 3–10 mg sample,
for 3–10% N, 5–20 mg sample,
for <3 N, 15–25 mg sample.

Solids and viscous liquids should be weighed into aluminium boats and about 0·3 g of copper(II) oxide added as described previously. Volatile samples are sealed in low melting glass capillary tubes. After sealing, they should be placed in an aluminium boat containing 0·3 g of copper(II) oxide as before. If the nitrogen content of the sample is less than about 0·5%, about 0·3 ml of air should be added to the nitrometer before the start of the test. The sample analysis should be carried out in exactly the same way as the blank at the end of the run. The temperature of the nitrometer and also the barometric pressure should be read.

The nitrogen content of the sample is calculated from the following equation:

$$\text{nitrogen} = \frac{PV}{K}\,\frac{44{\cdot}899}{\text{wt. of sample (mg)}}\ \text{wt\%}$$

where P = barometric pressure (mmHg) after correction for the vapour pressure of the water vapour over the potassium hydroxide;
V = volume of nitrogen (ml) after subtraction of the blank;
K = temperature of nitrometer (K).

If the instrument is to be shut down, all the furnaces should be switched off and the oxygen supply valve closed. The carbon dioxide flow-rate should be reduced to about 5 ml/minute. The nitrometer levelling bottle should be lowered and the spherical joint between the nitrometer and the combustion unit disconnected.

V. ACCURACY AND SOURCES OF ERROR

Before the instrument is put into routine operation, the accuracy of the temperature-indicating meters should be checked by means of a pyrometer. It is also advisable to check the flow-meter calibrations. In general the mechanical parts of the apparatus are very reliable. Some trouble may be experienced with the malfunctioning of the photoelectric sensing device due to particulate matter or froth on the surface of the potassium hydroxide. The admission cock can also be a source of trouble if it is not greased regularly. Leaks occur occasionally, but they will be found by the pressure drop test at the beginning of the day. They can usually be cured by tightening the compression nuts on the O-ring joints or by renewing the O-rings.

The performance of the instrument should be monitored by analyzing standard samples, i.e. samples whose nitrogen content has been obtained by calculation or by analysis using another method. A standard sample should be run several times a day. The blanks are fairly constant, particularly if the instrument is left on overnight, but should be checked at least once a day, and always after the oxygen or carbon dioxide cylinders have been changed. Loss of activity of the packings or channelling can lead to errors, usually positive. These will become apparent from the results obtained for the standard samples. It is useful to keep a log and to renew the packings after a certain number of samples have been run; the actual number will depend on the type and weight of the samples. Large samples can also lead to high results. The maximum sample size depends on the type of sample and also on the conditions of the furnace packings. When the additional copper oxide furnace H is used, the maximum sample size can vary between 25 and 100 mg. Negative results will be obtained if the combustion tube **B** becomes so full of oxidized boats that new samples cannot reach the high temperature zone.

The apparatus has been applied to a wide variety of samples both organic and inorganic. Merz[2, 3] has published figures showing the accuracy and precision of results obtained with the standard form of the apparatus for a series of organic compounds; they are shown in Table 1.

Merz also showed that the results obtained with a wide variety of fertilizers containing 10–30% nitrogen were in close agreement with the results obtained by the standard method.

When the method was applied to difficult samples such as hydrocarbon polymers containing about 1% nitrogen, the standard deviation was 0·03%.

Table 1. Results obtained for a series of organic compounds

Substance	% N (calculated)	% N found (mean)	ΔN	S
Acetanilide	10·36	10·36	0·00	0·10
1-Chloro-2,4-dinitrobenzene	13·82	13·75	−0·07	0·10
Azobenzene	15·37	15·36	−0·01	0·10
Copper phthalocyanine	19·46	19·40	−0·06	0·13
2,4-Dinitro-phenylhydrazine	28·28	28·26	−0·02	0·13
Urea	46·66	46·64	−0·02	0·08
Melamine	66·63	66·59	−0·04	0·08
Lupolen	0·20	0·24	+0·04	0·08
N,N,N′,N′-Tetramethylethylene-diamine	24·09	24·13	+0·04	0·09
Sulphanilic acid	8·09	8·08	−0·01	0·12
Nitrocellulose	12·61	12·56	−0·05	0·14

Note: ΔN = difference between mean and calculated value; S = standard deviation. 10 determinations made on each; sample weights 3–10 mg.

VI. REFERENCES

1. See e.g. G. Ingram, "Methods of Organic Elemental Analysis", Chapman and Hall, London, 1962.
2. W. Merz, *Z. Anal. Chem.* **237** (1968), 272.
3. W. Merz, *Amer. Lab.* **5** (12) (1973), 25.
4. T. L. Lunder, *Lab. Practice,* **23** (1974), 172.

6. MICROCOULOMETRIC METHODS FOR THE DETERMINATION OF SULPHUR, CHLORINE, NITROGEN AND INDIVIDUAL COMPOUNDS

F. C. A. KILLER

Esso Research Centre, Abingdon, Oxfordshire.

I. INTRODUCTION

During recent years microcoulometry has become widely accepted as a method for the determination of low concentrations of sulphur, chlorine and nitrogen in petroleum products and other organic and inorganic media. The major advantages of the method are its sensitivity, selectivity and speed, linearity of response and relative freedom from interferences. It is a true microtechnique in that it requires only microgram or even nanogram amounts of sample and can be applied to the analysis of gaseous, liquid or solid materials, alone or in combination with combustion, hydrogenolysis, selective adsorption and absorption, extraction, pyrolysis, catalytic and non-catalytic reactions and chromatographic separations to determine the total content of the above elements, or to distinguish between individual compounds or compound classes, containing these elements. The method has also found other uses, e.g. the determination of traces of water, but its possibilities have by no means been exhausted.

The early work of Liberti and Cartoni,[1] who in 1957 determined thiols in gasolines by coulometric titration of the individual compounds with silver ions after gas-chromatographic separation, passed more or less unnoticed. The potential of the technique was realised only when Coulson and co-workers[2, 3] introduced the microcoulometer in 1960 for the determination of organochlorine pesticide residues in foods, using it as a chlorine-specific gas-chromatographic detector. For this purpose the effluent from the gas chromatograph was led into a furnace where the organochlorine compounds were converted to hydrogen chloride by combustion in oxygen and titrated with electrogenerated silver. In 1961 Klaas[4] applied the same combination of gas-chromatographic separation, combustion and coulometric titration to the determination of sulphur compounds in naphthas, by titrating the sulphur dioxide formed in the combustion with electrogenerated bromine. Numerous publications followed,[5–17] describing the application of this combined technique to the determination of sulphur, halogen and phosphorus compounds in various materials such as liquid petroleum fractions, gases, pesticides and drugs.

The microcoulometer offered an attractive possibility of determining total sulphur (or chlorine) in materials such as light petroleum products by direct injection of the sample into the combustion tube, omitting the preliminary gas-chromatographic separation. It was soon realized that the combustion unit designed for the combustion of GC effluents was not suited for this purpose. Drushel[18] replaced the original combustion tube by a larger tube, suitable for the direct injection of samples at a temperature high enough to guarantee rapid and complete volatilization. Results obtained with this modified microcoulometric system for total sulphur in petroleum products were discussed by Drushel,[19] Killer and Underhill,[20] Moore and McNulty,[21] and others.

Wallace, Joyce and co-workers[22] introduced a reductive method for the determination of total sulphur, by reducing the sulphur compounds present in the sample over a platinum-on-alumina catalyst in a stream of hydrogen to hydrogen sulphide which was then titrated with silver ions. These authors, Killer[23] and Braier *et al.*[24] compared this method with the oxidative method for sulphur.

Hofstader[25] and Gunther and Barkley[26] were the first to apply combustion plus microcoulometry for the determination of total chlorine in samples containing organochlorine compounds.

Martin[27] pioneered the use of the microcoulometer for the determination of total nitrogen, after conversion of the nitrogen compounds to ammonia by catalytic hydrogenolysis over nickel, which was then titrated with electrogenerated hydrogen ions. Drushel[18, 19] and Moore and McNulty[28] proposed improvements of the system and Albert and co-workers[29] modified the reactor by adding another sample inlet tube which by-passed the hydrogenolysis reactor, thus permitting determination of total nitrogen and ammonia in aqueous samples such as waste water. Gouverneur and van der Craats [30] compared the microcoulometric method with other methods for the determination of nitrogen in organic materials.

Martin[27] also pioneered the application of the microcoulometer as a specific gas-chromatographic detector for nitrogen compounds and demonstrated the capabilities of the technique by determining the distribution of nitrogen compounds in catalytic cycle oil and shale naphtha. Albert[31] extended the use of this combination of methods to other petroleum fractions, and Drushel[32] combined microcoulometry with pyrolysis and gas chromatography to determine nitrogen compound classes in high-boiling petroleum fractions.

In the applications enumerated above, microcoulometry has been used primarily for the analysis of discrete samples. However, the term also applies to the continuous determination of trace quantities of gases and vapours in gaseous media by means of automatic coulometric analyzers which are based on the same null-balance coulometric principle as the laboratory microcoulometers for the analysis of discrete samples, by maintaining a small fixed concentration of titrant in the cell. Titrations of this type were first introduced by Shaffer, Briglio and Brockman(33) who, during the 1939–45 war, designed an instrument for the continuous determination of mustard gas in the atmosphere with electrogenerated bromine. Austin and co-workers(34) improved this instrument which has since been used extensively for the determination of oxidizable components such as hydrogen sulphide, thiols and sulphur dioxide in gas streams.(35, 36, 37) In fact, such continuous coulometric titrations have also been used in combination with gas chromatographs for the determination of discrete samples, with (4) and without (38) combustion.

Ryland and Tamele(39) reviewed the applications of microcoulometry to the determination of thiols, and Drushel(40) to the determination of organic sulphides. Several authors(41–45) have discussed the advantages and shortcomings of microcoulometry as a gas-chromatographic detector and compared it with other detectors.

Although microcoulometry was used first in the form of element-specific gas-chromatographic detectors for the analysis of discrete samples or in the form of continuous analyzers for trace contaminants in gaseous media, it is today considered primarily as a self-contained microanalytical technique for the determination of organically bound sulphur, chlorine and nitrogen, following an oxidative or reductive degradation step. In this capacity its sensitivity, selectivity and speed is hardly rivalled by other techniques. As a gas-chromatographic detector it has been replaced in some applications by other detectors, such as the flame-photometric detector(46) for sulphur and phosphorus which has greater sensitivity and speed response, the Coulson conductivity detector(47) which is preferred for the analysis of nitrogen compounds, and others.

Microcoulometry will therefore be discussed primarily in all its aspects as a method of elemental analysis, followed by its capabilities and scope in continuous trace analysis and in combination with other analytical techniques.

II. PRINCIPLES

A. Theory of coulometric titration

According to Faraday's law, the quantity of electrode reaction is directly proportional to the quantity of electricity expressed in coulombs (1 coulomb = 1 amp sec.), as measured by the time integral of the current I:

$$Q = \int_0^t I \, dt \tag{1}$$

If W is the weight in grams of material converted by Q coulombs, M the molecular or atomic weight of the substance, and n the number of electrons involved per mole of electrode reaction we have:

$$W = \frac{QM}{nF} \tag{2}$$

where F is the Faraday, i.e. the number of coulombs required for the transformation of one gram-equivalent of the substance. The value of F is approximately 96 500 coulomb per equivalent. The determination of substance transformed by measuring the quantity of electricity is the basis of coulometric analysis, and a quantitative overall titration reaction that proceeds with 100% current efficiency is essential for the successful application of the method.

The substance being determined may react directly at one of the electrodes (primary process) or it may react in solution with another substance, a coulometric intermediate, generated by an electrode reaction (secondary process). Microcoulometry is based on the latter case, since it employs titrants which are electrogenerated in or from the electrolyte. Diffusion provides a steady supply of reagent, thus maintaining a constant electrode potential which in turn guarantees 100% current efficiency.

B. Principles of microcoulometry

The high sensitivity and fast response of microcoulometers is achieved by combining the principle of null-balance coulometry with adequate cell design and electrolyte composition. A functional scheme of the system is shown in Fig. 1.

The reference electrode provides a constant EMF as a reference voltage, which is a direct function of the ion concentration on the electrode surface. It is usually separated, by means of a diffusion barrier, from the main cell cavity where the titration takes place without breaking the electrical contact with the sensor electrode which is located in the cell cavity, surrounded by the electrolyte. This indicator electrode pair provides a signal to the amplifier which is opposed by an external bias voltage, selected so that it corresponds to a low concentration (of the order of micromoles) of titrant in the electrolyte. When the two voltages are equal, the signal ΔE to the amplifier will be zero, the amplifier output will also be zero, and no current I will flow between

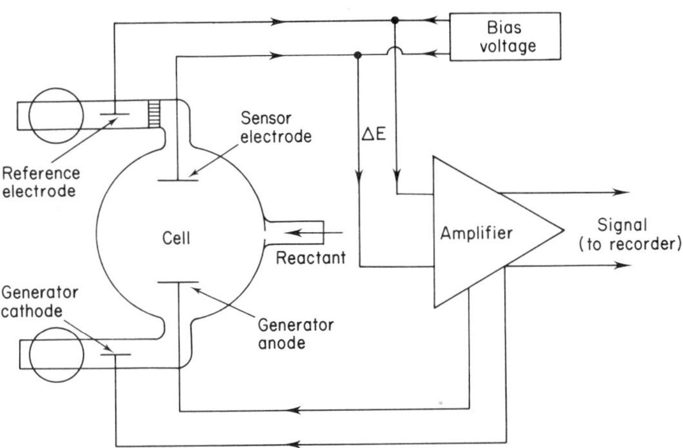

FIG. 1. Principle of microcoulometer.

the generator electrodes. The system is in balance:

$$E_{ref} + E_{sens} = E_{bias} \tag{3}$$

$$\Delta E = 0, \qquad I = 0, \quad \therefore \text{ system in balance}$$

Any substance entering the cell which reacts with the titrant ion will change the titrant ion concentration and hence the potential of the sensor electrode. As a result, the net output of the indicator electrode pair will not be equal to the bias potential and a signal ΔE different from zero will appear at the amplifier input. In response, the amplifier will supply an amplified voltage to the generator electrode pair and a current I corresponding to the magnitude of ΔE will flow in the titration cell, generating titrant ion. This process will continue until sufficient titrant ion has been generated to restore the sensor/reference voltage to the bias voltage, i.e. to restore the initial titrant ion concentration in the electrolyte:

$$E_{ref} + E_{sens} \neq E_{bias} \tag{4}$$

$$\Delta E \neq 0, \qquad I = f(\Delta E), \quad \therefore \text{ system generates.}$$

Thus, in contrast to classical "constant-current" coulometry, microcoulometry may be characterised as "variable-current" coulometry.

The microcoulometric titration process for a discrete sample is shown in Fig. 2 as it appears on the recorder chart. A straight line (base line) indicates that the microcoulometer is in balance, i.e. that the titrant concentration in the cell is constant and equal to the concentration corresponding to the bias potential.(48) When a substance which consumes the titrant ion enters the titration cell at time t_1, there is a sharp increase in generating current which

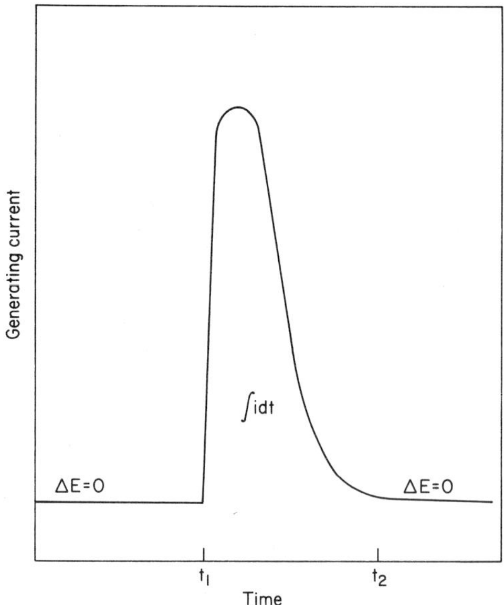

FIG. 2. Microcoulometric titration curve.

then levels off and returns at time t_2 to the initial low value, required to maintain the pre-set titrant ion concentration. The integral of the area under the curve is a measure of the current flow (in coulombs per unit time) consumed in the process and is proportional to the amount of substance titrated.

If the substance to be titrated is supplied to the cell in a continuous and constant stream, the generating current will again increase and remain at a new level as long as the supply of the titratable substance lasts. The height of this level above the preset bias-controlled generating current level is a function of the current flow consumed and is proportional to the concentration of the substance titrated.

To achieve maximum sensitivity and speed of response, the cell volume must be kept small and the titrant concentration low. The cell volume can be reduced, for instance, by removing those electrodes which do not participate directly in the titration (the reference electrode and the auxiliary generator electrode) from the cell cavity into side arms. Furthermore, the substance to be titrated must be brought into the cell cavity in the shortest possible way, to avoid unnecessary dilution and mixing which lead to peak broadening. If the titration is preceded by a pyrolysis step, as is the case in most applications of microcoulometry, the substance to be titrated is swept into the cell by the gas stream emerging from the pyrolysis tube. The gas stream itself provides agitation of the solution, but stirring of the electrolyte at constant speed is required to maintain a homogeneous solution, 100% current efficiency, and a defined diffusion layer thickness.

The role of electrolyte composition is best explained on the example of the titration of chloride with silver ions.[2] In this case the sensor electrode and the generator anode are made of silver, the reference is a silver–silver acetate electrode and the generator cathode is made of platinum. An aqueous acetic acid solution is used as the electrolyte instead of pure water to reduce the solubility of silver chloride, which is 10^{-5} mol/litre in water, but only 10^{-7} mol/litre in 80% acetic acid.

The generator anode and sensor electrode reactions in the cell are both

$$Ag^0 \rightleftharpoons Ag^+ + e^- \tag{5}$$

As described above, the microcoulometer is in balance when the potential difference between the reference and sensor electrodes is equal to the external bias voltage E_{bias}, which is chosen so that the silver ion concentration is approximately equal to the solubility of silver chloride. Under these conditions the chloride and silver ion concentrations are approximately equal so that the potential of the silver sensor electrode changes most rapidly for small changes in the amount of silver (or chloride) in the solution. If water is used as the electrolyte this occurs at 10^{-5} mol/litre, in 70% acetic acid at 10^{-7} mol/litre. Thus, the same amount of chloride entering the cell will represent in the latter case a greater change in silver concentration.

Cell volume has a similar effect on sensitivity. For instance, the addition of 5 nanograms (5×10^{-9} g) of silver to a microcoulometric cell which contains approximately 5 ml of electrolyte with 10^{-7} mol/litre of silver represents an increase of 1×10^{-8} mol/litre or 10%. If the cell contains 20 ml of the same electrolyte this addition would represent only an increase of 2·5%. Thus, cell volume affects not only sensitivity but also speed of response which is an important cell characteristic when the microcoulometer is to be used as a gas-chromatographic detector.

Table 1. Titration cell characteristics[2]

Cell volume	5 ml	20 ml
Time constant	2–5 sec	10–20 sec
Minimum detectable amount of chloride, g	2×10^{-9}	2×10^{-8}

C. Chemistry of combustion processes

In the combustion of organic compounds containing heteroatoms such as oxygen, sulphur, halogen, nitrogen or phosphorus, the following major reaction products will be formed in the presence of excess of oxygen.

Constituents	*Reaction products*
C, H, O	CO_2, H_2O
S	SO_2, SO_3
X (halogen except F)	HX, HXO, X_2
N	NO, NO_2
P	$P_2O_5(P_4O_{10})$

Carbon dioxide and water are the main reaction products of compounds containing only carbon, hydrogen and oxygen if the combustion is carried out under conditions warranting complete combustion. Neither of these reaction products affects the titrant ions used in microcoulometric cells. If conditions are not fulfilled, olefins and various aldehydes may be formed that react with iodine or bromine and will, therefore, interfere in the oxidative determination of sulphur.

Sulphur compounds are converted mainly to sulphur dioxide, but conversion never reaches 100%. It is mainly a function of temperature, but also depends on the partial pressure of oxygen and the design of the combustion tube. Sulphur dioxide may be oxidised to sulphur trioxide according to the equation:

$$SO_2 + \tfrac{1}{2}O_2 \rightleftharpoons SO_3 \tag{6}$$

The thermodynamic equilibrium constant K_p is given by:

$$K_p = \frac{p_{SO_2} \cdot p_O^{\frac{1}{2}}}{p_{SO_3}} \tag{7}$$

where p denotes the partial pressure of the respective gases. K_p is a function of the absolute temperature T according to:

$$\log K_p = 6{\cdot}38 \log T - 9480/T - 0{\cdot}0049T - 5{\cdot}41 \tag{8}$$

and increases from 0·7 at 800°C to 2·1 at 900° and 6·0 at 1000°.[49]

The theoretical recovery of SO_2 in equilibrium with O_2 and SO_3 is shown in Fig. 3 as function of temperature and oxygen pressure.[50] It can be seen that recovery will be the highest at a high temperature and a low partial pressure of oxygen. In practice, however, a rather high oxygen pressure is required to achieve a satisfactory combustion of the organic matrix, thus making the conversion to sulphur dioxide less than 100%. At an oxygen pressure of one atmosphere a temperature of 1412°C would be required to reach an SO_2 yield of 98%; at 900°C this value drops to about 75%. Practical considerations, such as the tendency of the quartz combustion tube to devitrify at temperatures above 1000°C and the fact that some sulphur compounds such as thiophene derivatives are difficult to decompose below

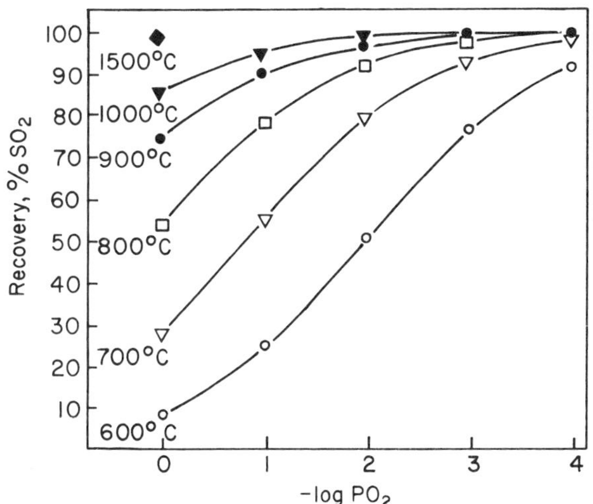

FIG. 3. Theoretical recovery of SO_2 in equilibrium with O_2 and SO_3 as a function of temperature (°C) and oxygen pressure (atm).[50]

800°C,[53] limit the combustion temperature to the range 800–1000°C. A temperature of 900°C is usually chosen for the combustion and a conversion of about 80% SO_2 is obtained which is in good agreement with the theoretical value. Such a recovery of sulphur dioxide is acceptable as long as it is kept constant, although it has been considered as a shortcoming of the oxidative method in comparison with the reductive method for sulphur which gives a 100% conversion of organic sulphur to hydrogen sulphide.[24]

The design of the combustion tube and the choice of operating conditions such as gas flow rates may also have an effect on the percentage of SO_2 conversion. These must be chosen so that the gas mixture leaves the combustion zone as fast as possible in order to "freeze" the equilibrium.

A computer study based on the free-energy minimization principle has been carried out by Cedergren[51] with the aim of predicting the equilibrium composition of the products resulting from combustion of a hydrocarbon sample containing sulphur and chlorine in an oxygen–argon atmosphere. Cyclohexane was chosen as the hydrocarbon and the system investigated can be represented as follows:

$$C_6H_{12} + xS_2 + yCl_2 + zO_2 + v\text{Ar} \rightarrow \text{Products} \tag{9}$$

The coefficients represent the number of moles of each substance present in the system: $x = 5 \times 10^{-5}$; $5 \times 10^{-5} < y < 0{\cdot}5$; $4{\cdot}5 < z < 18$; $v = 25$ and 250 mol. The total pressure was taken at 1 atm and the equilibrium composition of the system calculated at two temperatures, 1000 and 1300 K. The main constituents of the combustion products taken into consideration were: CO, CO_2, COS, H_2S, HCl, HClO, ClO, Cl_2, Cl, H_2O, SO_2, SO_3, H_2, O_2 and Ar.

Figure 4 shows the distribution of the sulphur species COS, H_2S, SO_2 and SO_3 as a function of moles of oxygen added to one mole of cyclohexane containing 5×10^{-5} mol sulphur. At oxygen concentrations above the stoichiometric ($z > 9$) practically all sulphur is present as SO_2 and SO_3. At 1300 K the mole fraction of SO_2 is near 100% and drops slightly with increasing amount of O_2 in the mixture. At 1000 K there is a significant drop in SO_2 (increase in SO_3) with increasing moles of O_2 per mole of hydrocarbon.

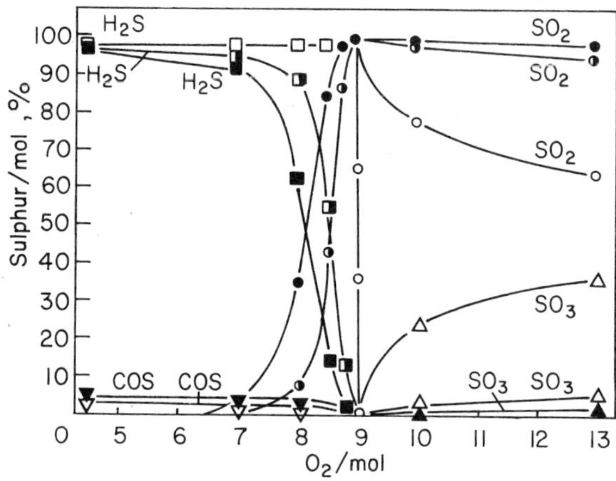

FIG. 4. The distribution of sulphur species COS, H_2S, SO_2 and SO_3 as a function of mol O_2 added to one mol of cyclohexane containing 5×10^{-5} mol of S_2: ○ SO_2, △ SO_3, □ H_2S, ▽ COS; (I) Unfilled symbols: 25 mol Ar, 1000 K, (II) partly filled symbols, 25 mol Ar, 1300 K, (III) filled symbols, 250 mol Ar, 1300 K.[51].

At oxygen concentrations below the stoichiometric ($z < 9$) the major sulphur species found is H_2S, with small amounts of COS also present. There is a sharp rise in the mole fraction of SO_2 (drop in H_2S) as the oxygen concentration approaches the stoichiometric value; for instance, at $z = 8{\cdot}75$ the equilibrium concentration is 86% SO_2, 12% H_2S and 2% COS for 25 mol Ar and 1300 K, and 98% SO_2, 2% H_2S and 0% COS for 250 mol Ar and 1300 K.

The unexpectedly high mole fraction of SO_2 formed with less than the stoichiometric amount of O_2 can be explained by taking into consideration the water gas equilibrium:

$$CO_{(g)} + H_2O_{(g)} \rightleftharpoons CO_{2(g)} + H_{2(g)} \tag{10}$$

At high temperatures the equilibrium is shifted to the left, thus providing an additional amount of oxygen (and hydrogen) in the system, according to:

$$H_2O_{(g)} = H_{2(g)} + \tfrac{1}{2}O_{2(g)} \tag{11}$$

This fact evidently favours the formation of SO_2.

Based on these facts, Cedergren comes to the interesting conclusion (51) that, in the oxidative determination of sulphur, carbon monoxide may be added to the system to reduce the p_{O_2} to a value which favours a high conversion to SO_2. The small amount of H_2S which is also formed under these conditions would also be titrated with iodine which serves as the titrant in the oxidative microcoulometric determination of sulphur. Both sulphur species reduce two equivalents of iodine.

These postulations seem to be supported by the results obtained by Cedergren[50] and Dixon[52] who diluted the gas mixture produced in a combustion carried out at relatively low temperature (700–900°C) with an inert gas at high temperature (1000°C). Thus, the partial pressure of oxygen was lowered and equilibrium obtained because of the high temperature used. The recovery of sulphur dioxide was close to 100% and the precision of the results better than that obtained with the usual (low temperature) combustion method. Theoretically, the dilution of the combustion gases with carbon monoxide also offers the possibility of suppressing the interference of some chlorine species in the determination of sulphur without affecting significantly the conversion of the organic sulphur to SO_2.[51]

Some investigators claim that the conversion of sulphur compounds to SO_2 is also dependent on the compound type and sample matrix.[22,24] Others have found that differences in conversion, attributable to different structures do not exceed 3% of the average value.[11,18,53] Marsh[54] obtained consistent results unaffected by sulphur type or sample matrix, although at a conversion level different from that expected for the working conditions used. The microcoulometric response obtained for various sulphur compounds over the range 5–200 nanogram of sulphur is shown in Fig. 5.

The fact that most workers have reported an SO_2 conversion unaffected by compound type and matrix leads to the assumption that the discrepancies may be due to effects related to the particular working conditions used, e.g. incomplete sample volatilization, incomplete combustion or non-equilibrium conditions.

The major product formed in the combustion of chlorine-containing organic compounds with excess of oxygen is hydrogen chloride. The primary combustion product is chlorine which reacts with water according to the scheme:

$$Cl_2 + H_2O \rightleftharpoons HCl + HClO \rightleftharpoons 2HCl + \tfrac{1}{2}O_2 \tag{12}$$

The hypochlorous acid formed as an intermediate decomposes quickly under

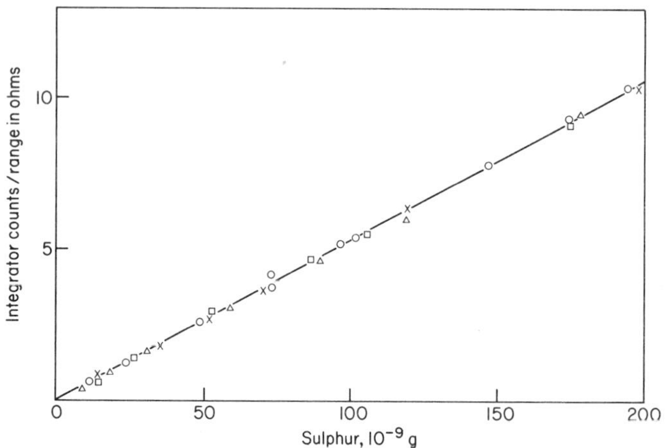

FIG. 5. Microcoulometric response of sulphur compounds as a function of the amount of sulphur injected: ○ Thiophene × propanethiol, □ dimethyl sulphide, △ diethyl sulphide (solutions in toluene).[20 23]

the conditions in the combustion tube, and the equilibrium is displaced to the right and results in an almost 100% recovery of chlorine as HCl. The role of water in this process has been pointed out by Coulson.[55]

The theoretical recovery of HCl in the modelled combustion of cyclohexane containing trace amounts of chlorine is shown in Fig. 6[51] as a function of moles of oxygen per mole of cyclohexane. Above the stoichiometric amount (>9 mol O_2/mol cyclohexane) there is a drop in HCl when the combustion is

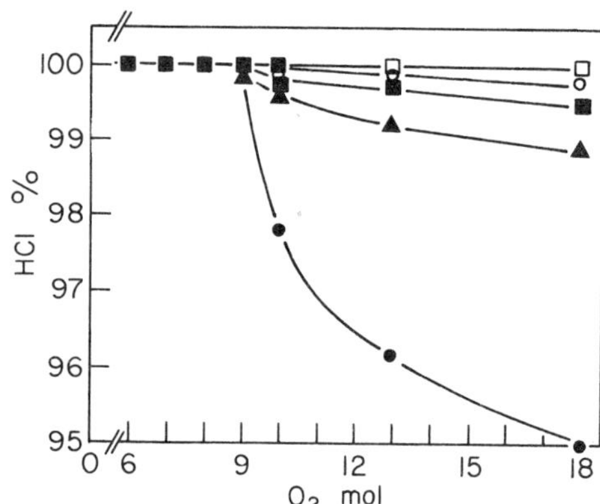

FIG. 6. The recovery of HCl as function of mol O_2 added to one mol of cyclohexane containing 5×10^{-3} mol Cl_2 (filled symbols) or 5×10^{-5} mol Cl_2 (unfilled symbols); ○ 25 mol Ar, 1000 K, △ 25 mol Ar, 1300 K, □ 250 mol Ar, 1300 K.[51]

carried out at a relatively low temperature (1000 K), which leads to the formation of other chlorine species, mainly of Cl_2 and HClO. An increase in combustion temperature to 1300 K suppresses this tendency almost completely.

The formation of Cl_2 and HClO is undesirable since it reduces the amount of titratable species reaching the cell in the coulometric determination of chlorine by titration with silver ions. It is also undesirable in the coulometric determination of sulphur by titration with iodine since Cl_2 and HClO oxidise iodide present in the electrolyte to iodine and thus affect the titrant concentrations in the cell. HCl has no effect on the titration of SO_2 with iodine. In practice it has been found that under the usual combustion conditions (900°C, excess oxygen) about 98% of the chlorine present in the sample is converted to HCl.

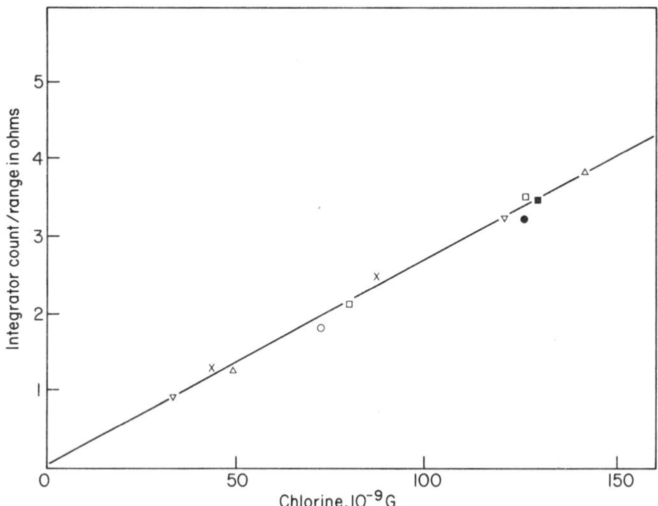

FIG. 7. Microcoulometric response of chlorine compounds as a function of the amount of chlorine injected: ○ 1-Chlorooctane, × 1,2-dichloroethane, □ chloroform, △ carbon tetrachloride, ▽ dichlorobenzene, ● 1-chloronaphthalene, ■ chloroaniline (solutions in toluene).

The conversion of chlorine compounds to HCl seems to be unaffected by compound type. The microcoulometric response of various chlorine compounds is a linear function of the amount of chlorine injected as shown by Drushel[18] and in Fig. 7 for the range 0–150 ng Cl.

The combustion of bromine-containing organic compounds leads to the formation of bromine which reacts with water to give a mixture of hydrogen bromide and hypobromous acid:

$$Br_2 + H_2O \rightleftharpoons HBr + HBrO \quad (13)$$

The aqueous solution of hypobromous acid is unstable and decomposes readily into bromine and water. Since Br_2 and HBrO do not react with silver ions in the titration cell, only about half of the bromine originally present in the sample is available as HBr for titration with silver ions. This is confirmed by data reported in the literature,[6, 56] showing a bromine recovery of about 50%, and by experience in the writer's laboratory in the analysis of motor gasolines containing dibromoethylene, where a conversion of 30–40% was obtained. Bromine present in the sample will also interfere seriously in the coulometric titration of SO_2 with iodine, since Br_2 and HBrO will oxidise iodide in the electrolyte to iodine as described above for chlorine.

The oxidative decomposition of iodine-containing organic compounds leads to the intermediate formation of hypoiodous acid which is very unstable in solution and decomposes to iodide and iodate:

$$3HIO \rightleftharpoons 2HI + HIO_3 \quad (14)$$

These then react together with the deposition of free iodine according to:

$$HIO_3 + 5HI \rightleftharpoons 3I_2 + 3H_2O \quad (15)$$

Thus, only a fraction of the iodine present in the sample will reach the titration cell in a form titratable with silver ions and the decomposition products other than HI will severely interfere in the coulometric titration of SO_2 with iodine.

Fluorine cannot be determined by titration with silver ions, nor does its presence interfere in the microcoulometric determination of sulphur.

The different behaviour of the individual halogens in combustion makes only chlorine amenable to microcoulometric determination and prevents the determination of total halogen in a sample if halogens other than chlorine are present. This difficulty may be overcome by subsequent reduction of the halogen species formed in combustion to hydrogen halides as described by Walisch and Jaenicke.[56]

Organic nitrogen compounds decompose in the combustion tube to give nitric oxide, NO, which with excess oxygen and on cooling forms nitrogen dioxide so that a mixture of the two oxides reaches the titration cell. If present in moderate concentrations, the oxides do not affect the coulometric titration with silver ions but they interfere with the determination of sulphur, since NO_2 liberates iodine from the potassium iodide present in the electrolyte.

Phosphorus pentoxide (P_4O_{10}) is presumably the main product formed in the combustion of organophosphorus compounds with excess oxygen. It has a very low volatility (it sublimes at 360°C) and is retained on the walls of the quartz combustion tube.[17]

D. Chemistry of hydrogenolysis processes

In the hydrogenolysis of organic compounds containing heteroatoms such as oxygen, sulphur, halogen, nitrogen or phosphorus, the following major reaction products will be formed in the presence of excess hydrogen and a catalyst:

Constituents	*Reaction products*
C, H, O	CH_4 +, H_2O
S	H_2S
X (halogen)	HX
N	NH_3, HCN
P	PH_3

The complete hydrogenolysis of compounds consisting only of carbon, hydrogen and oxygen leads to water and methane plus some higher homologs as the main reaction products. They do not interfere in the main applications of the reductive microcoulometric method, i.e. the determination of sulphur by titration of the H_2S formed with electrogenerated silver ions or the determination of nitrogen by titration of the ammonia formed with electrogenerated hydrogen ions. The hydrogen gas used in the hydrogenolysis must be humidified to prevent carbon formation and deposition on the catalyst at the relatively high temperatures in the catalyst zone (800–1150°C) which leads to decreased catalyst activity.

In the hydrogenolysis of organic sulphur compounds, the conversion of sulphur to H_2S is nearly stoichiometric (see Fig. 4) and depends very little on sample matrix, sulphur compound type and sulphur concentration.(22, 24) Aavik *et al.*(57) studied the conversion of sulphur compounds to H_2S (and of chlorine compounds to HCl) at various temperatures in the presence of platinum and palladium on various supports, and without a metal catalyst. A 100% conversion was obtained with 5% platinum on vitreous quartz (325 mesh) at 250°C, as well as with a quartz tube packed with quartz chips 60/80 mesh, and with a quartz capillary at 950°C.

Halogen compounds are converted to hydrogen halides in almost quantitative yields. However, the reductive method can be used for halogen determination only in the absence of sulphur compounds, since the H_2S formed also reacts with the silver titrant. On the other hand, interference of the hydrogen halides in the determination of sulphur can be eliminated by proper choice of the cell electrolyte.(22) This point will be discussed in more detail later. In the determination of nitrogen as ammonia by electrogenerated hydrogen ion, hydrogen halides (and any other acidic hydrogenolysis products) must be removed from the gas stream before it reaches the titration cell.

Ammonia is the main product of hydrogenolysis of organic nitrogen compounds. The conversion rate under the conditions usually applied for the reduction has been found to be 95–100%.[18] Values lower than this indicate catalyst deactivation, excessive coke formation or interference from strongly acidic components (HCl) in the hydrogenolysis. In general, the life of the catalyst decreases as the boiling point of the samples increases.[27] The microcoulometer response is the same for different nitrogen compounds, even for some heterocyclic compounds that are difficult to decompose by Kjeldahl digestion, as shown in Fig. 8, where the response for different compounds is plotted against nitrogen concentration.[18]

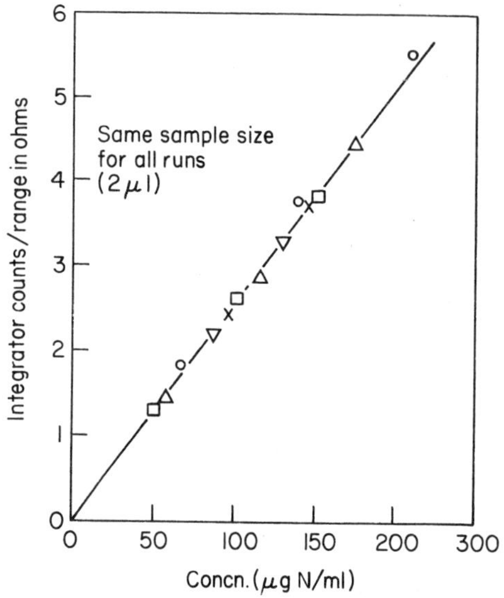

FIG. 8. Microcoulometric response of nitrogen compounds as function of nitrogen concentration: ○ Pyridine, □ quinoline, × benzoquinoline, △ indole, ▽ carbazole.[18]

Nitrogen compounds in the sample also lead to the formation of HCN which will be titrated with silver ions under the conditions selected for sulphur determination and will thus interfere in the determination. The amount of hydrogen cyanide formed is proportional to the carbon-to-nitrogen ratio in the sample. Humidification of the hydrogen gas and an increase in pyrolysis temperature reduce HCN formation.[22]

The reduction of organic phosphorus compounds with molecular hydrogen at elevated temperatures yields phosphine PH_3. Aavik and co-workers[58] studied the conversion of some di- and tri-alkyl phosphates and phosphites at temperatures below 400°C in the presence of a catalyst and at 900–1000°C in an empty quartz tube or a tube packed with quartz chips and obtained complete conversion at 350–370°C with 5% platinum on quartz chips, as well as at 900°C in a tube packed with quartz chips alone.

Phosphine precipitates silver ions. This approach can thus be used for the microcoulometric determination of phosphorus in organic phosphates.[5, 6, 17] H_2S and HCl when present in the effluent from the reduction tube interfere in the titration since they also react with silver ion. This interference can be overcome by passing gases over a bed of alumina which retains hydrogen sulphide and HCl but passes the phosphine. PH_3 and H_2S in the presence of one each other can be measured by separating them on a short gas chromatographic column packed with silica gel inserted between the reduction tube and the titration cell. HCl is retained irreversibly by this column. If the temperature of the reduction tube is lowered to 700°C the phosphate moiety does not yield PH_3, but sulphur atoms bonded to phosphorus are reduced to H_2S at this temperature. This offers the possibility of distinguishing between sulphur atoms bonded to phosphorus and those linked to other atoms.[5]

III. APPARATUS

The basic microcoulometric system as used for elemental analysis consists of an oxidative or reductive decomposition unit, the microcoulometer which incorporates the electronic circuitry, and the coulometric titration cell. A device for the recording of the titration curve and integration of the current-time relationship is usually added to the system. The Dohrmann microcoulometric system is shown in Fig. 9, consisting of the furnace unit (bottom right), the microcoulometer with an incorporated integrator placed on top of the furnace, and the cell cabinet, which houses the titration cell and the magnetic stirrer. A block scheme of the system is shown in Fig. 10. The sample is injected into the pyrolysis tube where it undergoes oxidative or reductive degradation. The products formed are swept by the stream of carrier and reactant gases into the titration cell, where the species formed from the sulphur, chlorine or nitrogen compounds in the sample are titrated with the appropriate electrogenerated titrant. The electric signal generated by this process is then processed by the microcoulometer, displayed on a stripchart recorder and/or evaluated quantitatively by means of an integrator. The processes occurring in the combustion or hydrogenolysis of the sample have been discussed in the previous chapter, and the electrode processes involved in the titration of the species formed will be described when discussing the procedures for the determination of sulphur, chlorine and nitrogen respectively.

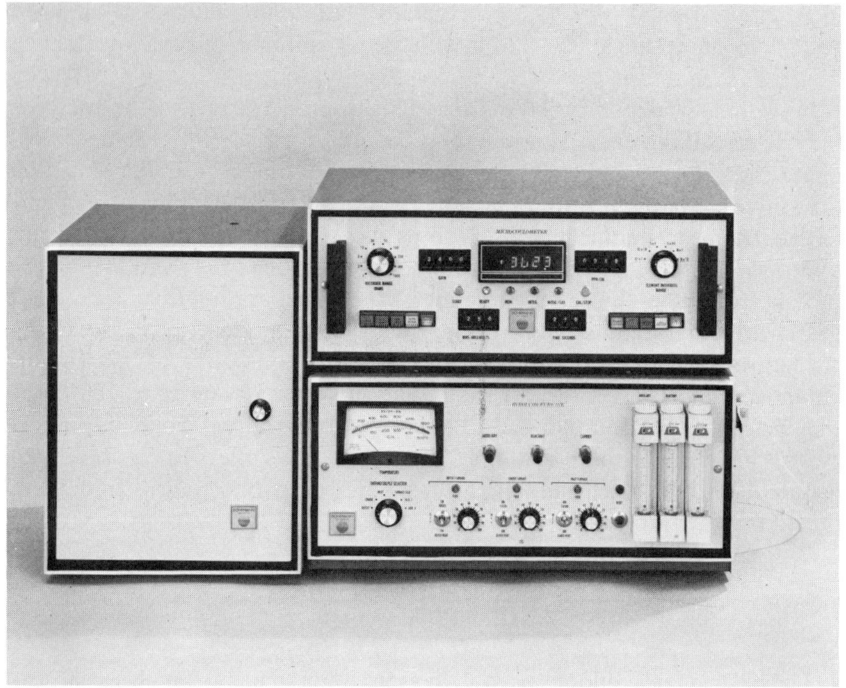

FIG. 9. Dohrmann microcoulometric system, consisting of furnace (bottom right), microcoulometer (top right) and titration cell (in cell housing, left). (Courtesy Dohrmann Envirotech, Mountain View, Cal.)

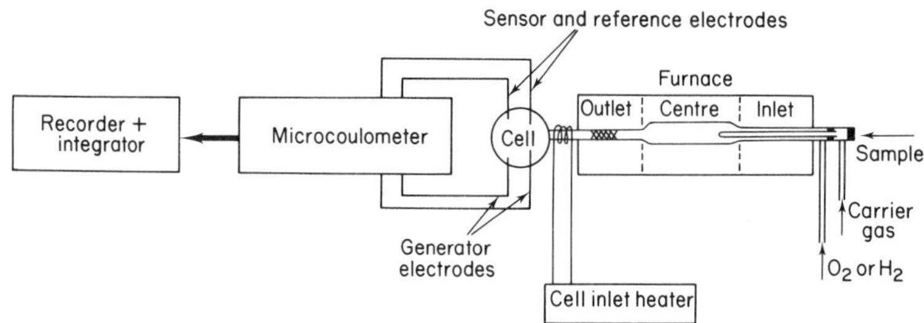

FIG. 10. Block scheme of microcoulometric system for elemental analysis.

A. Pyrolysis tubes and furnace

A typical tube for the oxidative determination of sulphur and chlorine is shown in Fig. 11. The tube is made of quartz and is used without a packing, except for a small plug of quartz wool in the outlet section. The sample

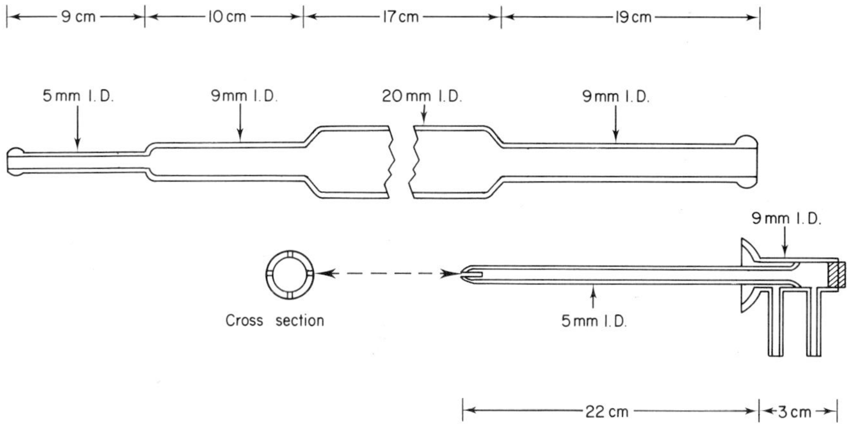

FIG. 11. Pyrolysis tube for the microcoulometric determination of sulphur or chlorine by the oxidative method.

injected through the silicone rubber septum is vaporised in the inlet section of the tube in an inert gas such as nitrogen, argon or helium, which enters the tube through the side arm near the septum. The vaporized sample is then swept by the gas through the narrow tube into the combustion zone and is burned in oxygen which is supplied to the combustion zone through the other side arm. Combustion takes place at the tip of the narrow tube which must be designed so as to provide intimate mixing of the sample vapours with oxygen and to prevent flashbacks which would lead to incomplete combustion and the formation of carbon deposits in the inlet section of the tube. Complete and reproducible combustion of the sample must be achieved and equilibrium conditions reached before the reaction products leave the combustion tube. The quartz wool plug in the outlet section of the tube will retain soot particles formed should incomplete combustion have occurred due to overloading of the tube capacity by too large a sample, too fast a sample injection, inadequate gas flow rates, etc., and thus assist in preventing contamination of the coulometric titration cell. The furnace temperatures and gas flow rates used will be discussed under the procedures for the determination of the individual elements.

Tubes similar to that shown in Fig. 11 have been used by other workers.[16, 20, 22, 52, 56, 59]

A tube of different design, such as the one shown in Fig. 12, must be used for the reductive determination of sulphur and nitrogen. It is also made of

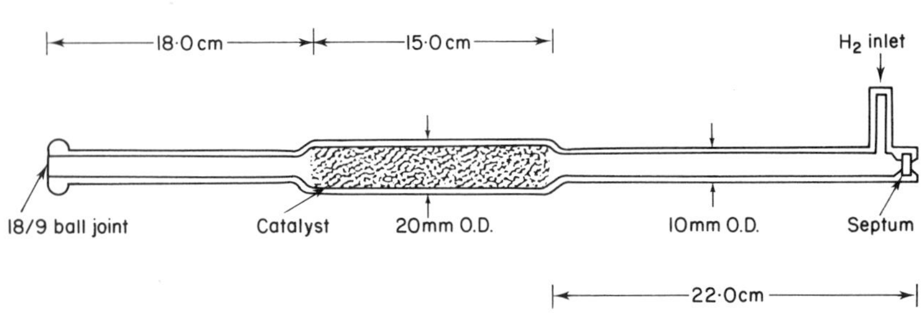

FIG. 12. Pyrolysis tube for the microcoulometric determination of sulphur or nitrogen by the reductive method.

quartz. The central wide section of the tube is packed with the appropriate catalyst (platinum-on-alundum for sulphur, nickel shot for nitrogen). The sample is injected through the silicone rubber septum and is vaporized in the inlet section of the tube in hydrogen, which in this method serves both as reactant and carrier gas and is admitted into the tube through the side arm near the septum. Hydrogenolysis of the sample takes place in the hot catalyst bed and the reaction products are swept into the titration cell where they are titrated appropriately. To achieve complete and reproducible hydrogenolysis of the sample, catalyst activity must be maintained at a high and constant level. Prime attention is paid to prevent the formation of coke deposits on the catalyst which severely affect its performance, and to avoid losses of sample before it reaches the catalyst. This problem is particularly acute in the analysis of high-boiling materials.

Tubes of similar design to that shown in Fig. 12 have also been used.[19, 22, 28, 61, 62] The furnace temperatures and gas flow rates used will be discussed under the procedures for the determination of the individual elements.

In the determination of nitrogen, acidic gases such as HCl, formed in the hydrogenolysis of chlorine-containing compounds, must be removed from the gas stream before they reach the cell because they interfere in the titration of ammonia formed from the nitrogen in the sample. Rhodes and Hopkins[64] absorbed the acidic gases in a packing of ascarite, placed between the exit of the pyrolysis tube and the capillary inlet to the titration cell, i.e. at room temperature. This lead to absorption of water which quickly saturated the scrubber packing and required frequent changing. To avoid this, the scrubber tube was inserted into the outlet section of the pyrolysis tube, kept at 300°C, and the ascarite packing was replaced with an inert support (alundum) moistened with sodium hydroxide. Such a scrubber tube, made of quartz, is shown in Fig. 13.

Drushel[19] prefers a scrubber tube made of stainless steel as it is less attacked by the hot alkaline packing than the quartz tube. Scrubber tubes

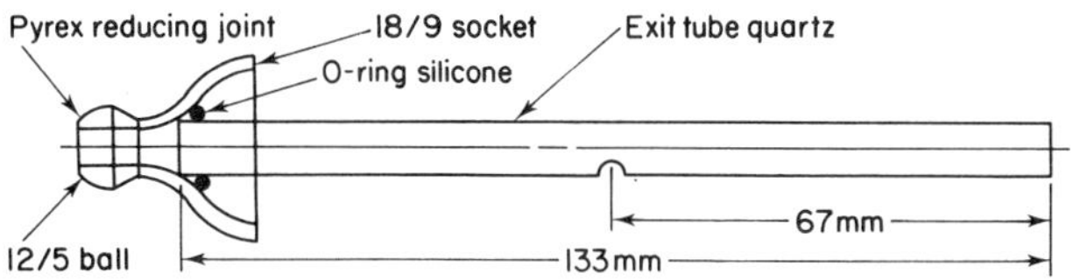

FIG. 13. Scrubber tube.[63]

packed with calcium, magnesium and strontium oxides and potassium carbonate have also been used.

The combination of microcoulometry with gas chromatography requires pyrolysis tubes with inlet sections adapted to receive the effluent from the gas chromatograph.[27,32,60] Tubes suitable for all the applications discussed in this chapter are now available commercially,* including universal tubes for combustion and hydrogenolysis, which have been designed to accept interchangeable sample inlets for gases, liquids and solids (COT and CRT tubes).[70]

Furnaces for microcoulometric analysis must be designed so that they can accommodate the combustion tubes described above. They must be able to maintain a preselected temperature in the range 400–1100°C to the nearest 5°C. The furnace should be divided into three zones of equal width, so that the temperatures in each zone can be varied independently within the above range. The heating circuits should include trip-out facilities to prevent temperature rises above 1100°C, in case of temperature control failure, which would damage the pyrolysis tube. Adequate furnaces are available commercially.†

B. Microcoulometer

The most widely used microcoulometer is the Dohrmann instrument originally developed by Coulson and Cavanagh.[2] A simplified functional scheme of the C-300 model which also includes an integrator section is shown in Fig. 14.

The microcoulometer section is based on fully transistorized switching circuits. The voltage of the sensor/reference electrode pair is sampled intermittently by means of the transistor switches (choppers) S_1 and S_2 and compared to the bias voltage V_1. When the microcoulometer is in balance the resulting output is zero. Any potential created when this balance is disturbed is amplified by the multistage amplifier system whose gain can be

* From Dohrmann Envirotech, Mountain View, California.

† e.g. From Dohrmann Envirotech, Mountain View, Cal. as a separate unit (S-300 or S-350 Furnace) or as part of integral analyzers.

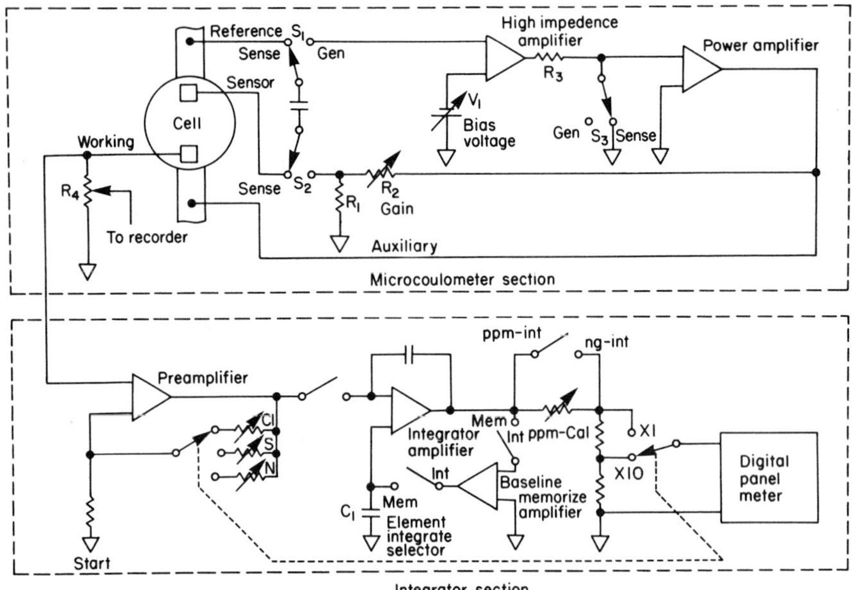

FIG. 14. Functional scheme of the Dohrmann C-300 microcoulometer.[65] S_1–S_3 transistor switches, R_1–R_4 resistors, V_1 bias potential, C_1 capacitor.

selected by varying the gain control R_2. The amplified output is applied to the generator electrode pair which generates titrant ion until the voltage of the sensor reference electrode pair is again equal to the bias voltage V_1. The transistor switches S_1 and S_2 are synchronized with S_3 so that no generating voltage is applied when the sensor/reference voltage is sampled. The titrant generation current flows through resistor R_4 creating a voltage signal which is available to the integrator section as well as to an optional recorder.

In the integrator section, the baseline memory circuit stores all off-balance voltages on capacitor C_1. As the reaction in the titration cell proceeds, the microcoulometer signal is simultaneously integrated for a preset period of time relative to the baseline memorized at the beginning of each integration cycle. The integrator output can be scaled so that the result obtained is displayed digitally in nanograms or parts per million of the element analysed.[65]

The Model 16300 LKB coulometric analyzer[66] is another microcoulometer of high sensitivity,* with a built-in integrator. The instrument can also be used for controlled-current coulometry. Some performance characteristics of the two microcoulometers are given in Table 2.

The Antek nanocoulometer† uses a pulsed current to restore the initial

* LKB-Producter AB, S 161 25, Bromma, Sweden.

† Antek Instruments, Inc., Houston, Texas.

Table 2. Performance characteristics of microcoulometers

	Dohrmann[65]	LKB[66]
Generator current, max.	±1·5 A	±0·6 A
Output voltage (DC), max.	±50 V	±40 V
Bias voltage control	0–1 V in 1-mV steps	pH 0–14(+700–800 mV) 0·01 pH or 1 mV scale div.
Gain control	0 to 10 000 in steps of 1	1:10 000 ca. 40% per step
Amplification factor	—	25 μA–200 mA per mV
Titration speed, max.	—	360 μeq/min

titrant concentration. The amount of current supplied with each pulse can be varied, thus allowing the analyst to select the number of counts for a given weight of ions titrated (from 0·1 to 4 ng per count). The manufacturers claim superior sensitivity for their instrument.

Some investigators have built microcoulometers to their own design.[56, 67] Of particular interest is the microcoulometer by de Groot *et al.*,[49] designed on the principles described by Krijgsman *et al.*,[68, 69] since it is built of commercially available components and instruments. The electronic scheme of the microcoulometer is shown in Fig. 15. The amplification factor of the coulometer can be varied from 0 to 75 μA per mV input signal.

Several microcoulometers are available commercially, designed specially for the continuous titration of sulphur species in gaseous media which can be oxidized with bromine or iodine. They are built as portable instruments or for use in the field, but may be adapted for the analysis of discrete samples.[4]

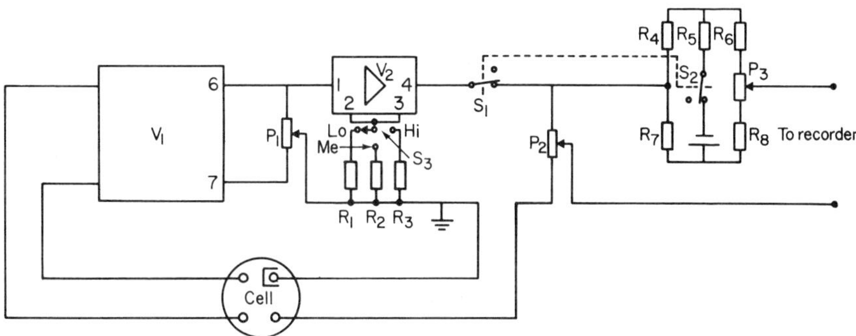

Fig. 15. Electronic scheme of microcoulometer by de Groot *et al.*:[49] V_1 Type 47 "Messzusatz", Knick. V_2 Type Hv d.c. amplifier, Knick, S_1, S_2 function switches, S_3 gain switch, P_1 amplifier gain potentiometer, P_2 recorder range potentiometer, P_3 recorder zero potentiometer. R_1–R_7, resistors, B mercury battery.

The Titrilog,* the Barton Titrator† and the Philips SO_2 Monitor‡ fall in this category. The functional scheme of the Titrilog is shown in Fig. 16. The gaseous sample is drawn by means of a pump into the inner compartment A which houses the indicator electrode 2 and the generator anode 1 where it is titrated with the electrogenerated titrant. Some of the electrolyte overflows through the overflow tube and returns to the stock electrolyte passing the charcoal filter. The flow rate of the gas must be adjusted to suit the concentration range of the substance being determined. The titrant (bromine) is regenerated against a reference battery (bias) voltage according to the null-balance coulometric principle.

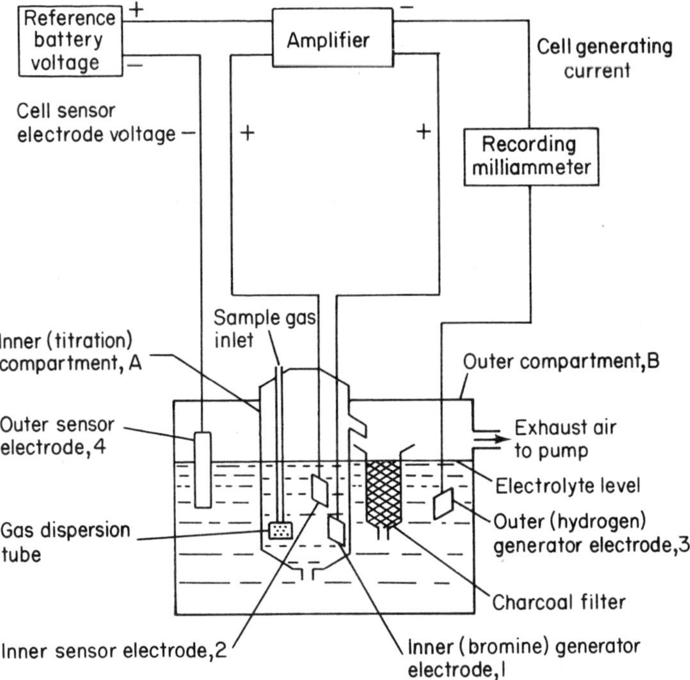

FIG. 16. Functional scheme of Titrilog continuous coulometric analyser.[71]

C. Titration cells

The most widely used titration cells for microcoulometry are those developed from the original design of Coulson and Cavanagh.[2] The cell for the

* Consolidated Electronics Corp., Pasadena, Cal.
† Model 286 Titrator, ITT Barton Instrument Co., Monterey Park, Cal.
‡ Model PW 9700 SO_2 Monitor, Philips Inc., Eindhoven, Netherlands.

oxidative determination of sulphur is shown in Fig. 17. It has a very small cell volume (ca. 5 ml) and thus fulfils the requirements of sensitivity and speed of response. The reference electrode 4 and the auxiliary generator electrode 10 are located in separate side arms. The sensor electrode 6 and the generator

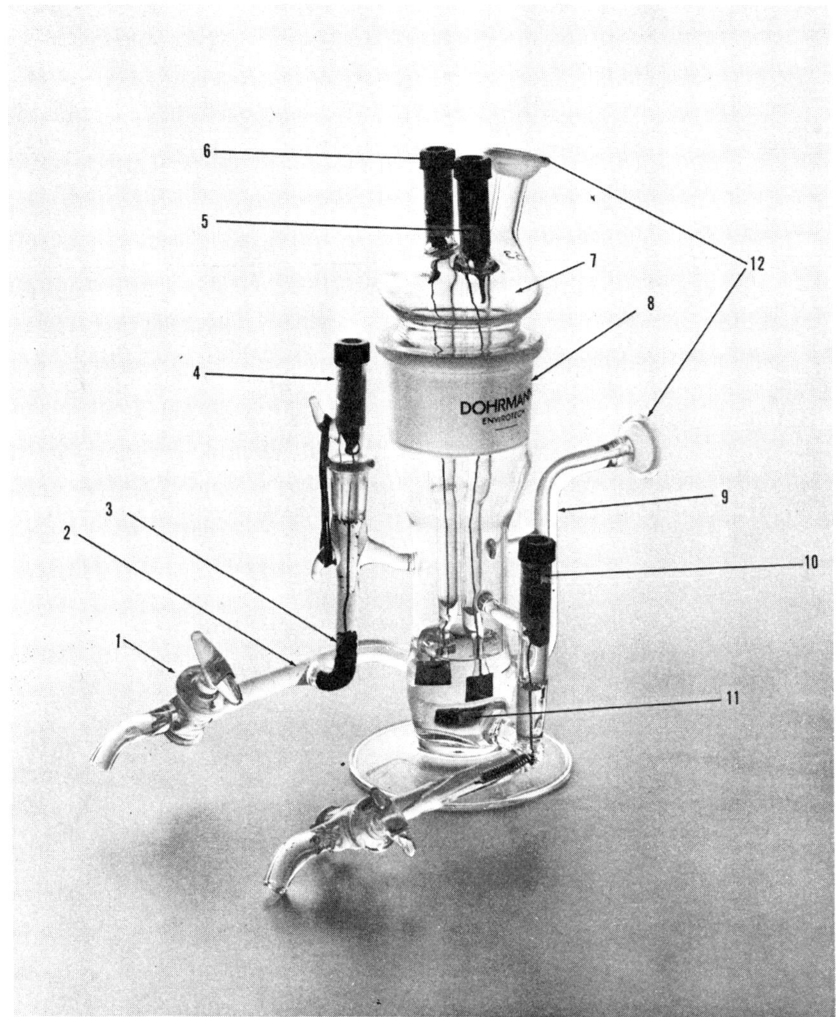

FIG. 17. Titration cell for the oxidative determination of sulphur (Model T-300-P, Dohrmann): (1) Side arm stopcock, (2) reference side arm, (3) reference electrode packing, (4) reference electrode, (5) generator anode (working electrode), (6) sensor electrode, (7) cell cap, (8) cell body, (9) capillary inlet, (10) generator cathode (auxiliary electrode), (11) stirring bar, (12) ball-joint sockets.[72]

anode (working electrode) 5 are mounted in the cap 7 so that they can be removed for easy cleaning. The stopcocks 1 facilitate flushing of the side arms with fresh electrolyte. The gases from the combustion tube enter the cell through the capillary inlet 9 which has a constriction at the point where it enters the cell cavity to break the gases up into small bubbles. The magnetic stirrer 11 provides rapid and thorough contact between the reacting species and the titrant and ensures a uniform diffusion layer. Two glass frit plugs in the reference side arm serve as diffusion barriers.

The same cell but with electrodes made of different materials serves for the oxidative determination of chlorine. A somewhat different cell with one more side arm and one more capillary gas inlet, again with different electrodes, is used for the reductive determination of sulphur and nitrogen. The two basic cell designs (in top view) and the electrode materials for the different applications are given in Fig. 18.

Adams *et al.*[(73)] converted the T-300-P cell to a bromine cell, by replacing the iodine reference electrode with one made up of mercury–mercury(I) bromide paste. The electrolyte is a solution of 0·72 g of potassium bromide and 6·0 ml of sulphuric acid per litre. The authors claim the sensitivity of their cell to be 30 times greater than that of the original iodine cell.

Cedergren and Johansson[(74)] used a cell with a rotating generator anode for the microcoulometric determination of chloride. The cell which is shown in Fig. 19 is made up of commercially available components, and is particu-

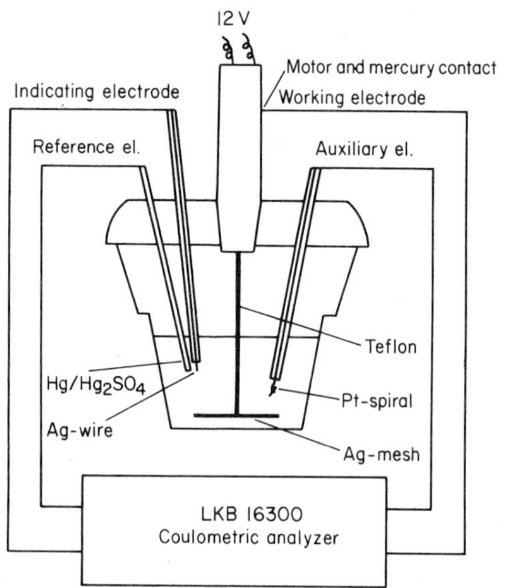

FIG. 19. Titration cell for chlorine with rotating generator anode.[(74)]

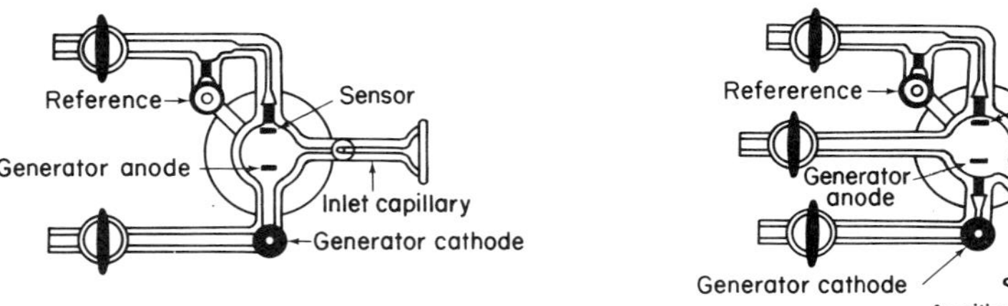

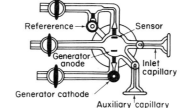

FIG. 18. Titration cells for the microcoulometric determination of sulphur, chlorine and nitrogen.[48]

Element	Sulphur	Chlorine	Sulphur	Nitrogen
Mode	oxidative	oxidative	reductive	reductive
Cell model*	T-300-P	T-300-S	T-400-S	T-400-H
Electrodes:				
Sensor	platinum	silver	silver	platinum (black)
reference	Pt°/I_3^- satd.	Ag°/Ag acetate satd.	Hg°/HgO satd.	$Pb/PbSO_4$ satd.
gen. anode	platinum	silver	silver	platinum
gen. cathode	platinum	platinum	platinum	platinum
Electrolyte:	0·05% potassium iodide 0·50% acetic acid 0·06% sodium azide	70% acetic acid	0·3 M ammonium hydroxide 0·1 M sodium acetate	1·0% sodium sulphate

larly suited for the direct introduction of liquid samples into the cell electrolyte. A silver wire is used as the indicating (sensor) electrode and a Radiometer K601 mercury–mercury(I) sulphate electrode as the reference. The rotating generator anode is made of silver gauze, fastened to the LKB rotating electrode assembly. A platinum spiral serves as the generator cathode. The titration vessel is a Metrohm EA 880–20 vessel with a volume of 20 ml, the electrolyte was 75% acetic acid.

Microcoulometric cells are usually sensitive to electrical disturbances from the surrounding medium. It is therefore advisable to protect them from capacitance effects by placing them into a grounded Faraday cage, made of wire gauze or sheet metal, or simply by wrapping them with grounded aluminium foil. Silver cells for the determination of chlorine are sensitive to light and must be protected by placing into a light-tight box or by painting them with black paint.

Titration cells made of glazed ceramic are also available.* They offer the advantages of greater ruggedness, no interference from incident light and easy direct sample injection into the cell. The electrodes are permanently fused onto the cell walls.

Other cell designs have also been proposed, e.g. by Liberti and Cartoni[(1)] for the titration of thiols with silver, and by Walisch and Jaenicke[(57)] for the titration of halogen with silver.

The cells used in the continuous coulometric analyzers are designed to suit the purpose of the particular analyzer. In this case, speed of response is not a critical parameter; hence the cells are relatively large and designed so as to remain operational over longer periods of time. A typical cell of this type is that used in the Titrilog analyzer (Fig. 16). The cell consists of an inner compartment in which the reactive portion of the sample is absorbed in the electrolyte and in which the titration takes place, and a larger volume surrounding the inner cell which serves as an electrolyte reservoir. The two vessels are connected by perforations in the bottom of the inner cell and an overflow near the top.

The electrodes consist of a sensor electrode made of platinum and a conventional calomel reference electrode, and of the generator electrode pair, also made of platinum. The sensor electrode and generator anode are placed in the inner cell, the reference electrode and the generator cathode in the outer reservoir. The electrolyte is a solution 3 M in sulphuric acid and 0·05 M in potassium bromide. As air is drawn into the inner cell through the sample intake tube, the air bubbles streaming through the cell act as an air lift; electrolyte from the inner cell flows through the overflow tube into the

* Antek Instruments Inc., Houston, Texas.

reservoir and is replaced by electrolyte, freshly filtered through a bed of charcoal granules on the bottom of the reservoir. The reactant in the air entering the cell is titrated with electrogenerated bromine and the generating current used in the usual way to establish the reactant concentration.

A magnetic stirrer is always required to agitate the cell electrolyte. An adequate stirrer must not produce electrical fields that may disturb the electrical conditions in the titration cell and must not create heat. The latter is frequently the case with stirrers which have the speed regulator incorporated in the stirrer housing. The electrochemical processes taking place in the cell are very sensitive to changes in temperature. For instance, McNulty and Hoppe[(48)] have calculated that, in the case of a lead/lead sulphate–hydrogen electrode pair, a mere increase in cell temperature of 1°C produces a shift of 0·8 mV in electrode potential which manifests itself as a baseline shift which becomes the more pronounced the higher the attenuation (range ohms).[(75)]

D. Recorder and integrator

Since the concentration of the element determined is a function of the current-time integral, peak area measured with an integrator rather than peak shape is essential for quantitative analysis by microcoulometry. However, it is advisable also to display the sample peak on a recorder to ensure that it has the correct shape, i.e. that it neither tails nor overshoots. For this purpose, a 1-mV full scale potentiometric recorder is recommended with a response of 1 second and a chart speed of about 1–2 cm/min. The input impedance of the recorder should be greater than 200 kohms.[(65)] Peak areas can be integrated satisfactorily by means of a Disc integrator fitted to the recorder; more sophisticated electronic integrators may be used instead.

E. Syringes

Liquid samples that can be completely vaporized in the inlet section of the pyrolysis tube are best injected by means of a microsyringe. The syringe volume must be chosen to suit the sample volume which is usually 1–20 microlitres. Needles of adequate length (about 10 cm) must be used to reach well into the hot inlet section of the combustion tube, thus avoiding flashbacks and condensation of the sample in cold parts of the inlet.

Microsyringes with the sample in the needle may also be used, usually for smaller volumes (1–5 μl). Compared with the normal syringes which hold the sample in the syringe barrel, these syringes have the advantage that the complete sample is injected, while with the normal syringe a certain portion of the sample remains in the cold part of the needle and has to be subtracted from the sample volume measured. On the other hand, air bubbles that may have formed when filling the syringe needle remain undetected, and great care must be taken to avoid coke deposits in the needle since these will reduce the nominal sample volume.

Multiple injection syringes have been recommended for the injection of larger samples (up to 50 μl), which use a repeating dispenser (such as the Hamilton Model PB 600–1), for the repeated injection of small amounts of sample (0·2–1 μl) at a controlled rate. These injections tend to produce flash combustions which, for instance in oxidative sulphur determination, disturb the equilibrium between the SO_2 formed in the combustion and the SO_2 adsorbed on the tube walls, thus affecting the repeatability of the results. For the injection of large samples at a controlled rate, the writer prefers the slow but continuous injection of the sample by means of a cranking device, such as the one described by Drushel,[76] or with a syringe pump.

Gas samples (usually 1–10 ml) can be injected with special gas-tight syringes. Shorter needles (about 6 cm) are adequate. Samples larger than 1 ml must be injected slowly, at a carefully controlled rate, so as not to exceed the combustion tube capacity and cause carbonization.

F. Sample inlets

1. *For solids and heavy liquids*

Solids and liquids that are not completely vaporized in the inlet section of the pyrolysis tube when injected with a syringe, may be introduced with a

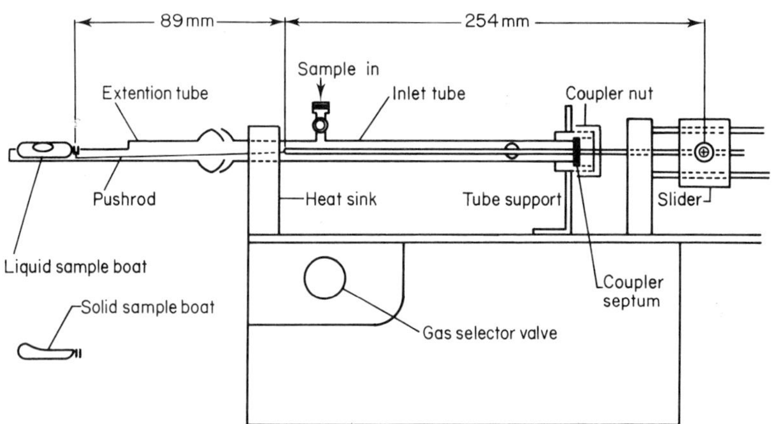

FIG. 20. Scheme of boat sample inlet (SBI Single Boat Inlet, Dohrmann).[77]

boat. Such a boat sample inlet is available from Dohrmann Envirotech and can be attached directly to the universal Dohrmann pyrolysis tubes (COT for oxidative and CRT for reductive analysis). A scheme of the inlet system is shown in Fig. 20.

The system is intended primarily for the analysis of heavy liquids and solids in the form of solutions in volatile solvents. For this purpose the boat (quartz boats are preferred to ceramic ones) is retracted by means of a platinum push rod to a position beneath the sample inlet. The sample (or sample solution) is then injected into the boat by means of a syringe through the septum in the sidearm. The boat with the sample is pushed towards the heated inlet section of the pyrolysis tube to a position where the temperature will give a uniform evaporation of the solvent (and volatile components of the sample). Too fast an evaporation rate must be avoided since it may overload the pyrolysis tube or cause condensation in the cooler parts of the inlet. When the evaporation is completed (usually after 20–60 sec) the boat is pushed into the hot inlet to a position where the sample is pyrolyzed for ca. one minute. The pyrolysis products are swept into the titration cell and titrated in the usual way to produce a peak. The boat is then withdrawn to the heat sink, allowed to cool for ca. 30 seconds, and returned to the initial position. It is now ready for the next sample.

The boat inlet is equipped with a gas selector (four-port) valve which makes it possible to replace the carrier gas with oxygen and thus to ensure complete combustion. Liquid sample sizes of up to 100 μl or solids up to 10 mg can be analyzed with this inlet system. It can also be applied for the direct introduction of solids, but then the system becomes tedious, since the boat has to be removed from the inlet system each time to weigh the sample. In this case a multiple boat inlet (also available from Dohrmann) which can take up to 17 boats is the better choice.[78]

The use of a quartz boat instead of a syringe with a metal needle can be recommended for the analysis of samples that contain very low concentrations of the element determined, since it avoids the appearance of the so-called "needle peaks". This problem will be discussed later.

2. *For gases and LPG*

A slide valve can be used instead of a syringe to introduce gaseous samples into the pyrolysis tube. Such a device is available from Dohrmann Envirotech with a flow system all made of glass and Teflon to reduce adsorption of sample components such as sulphur compounds on, and avoid reaction with, metal surfaces. The flow scheme of the device is shown in Fig. 21.

To introduce a gas sample into the pyrolysis tube, the slide valve (gas sample slider) is pushed into the position where the carrier gas flows straight through the slide valve and into the pyrolysis tube. With the valve A on the gas sampling line fully open, a slow flow about 100 ml/min of the sample is adjusted by means of the valve on the pressurized sample container. The gas sample now flows through the other arm of the slide valve, the sample loop, the flow meter, and to the vent. The sample loop volume is usually 5 or 20 ml, but any other volume can be chosen, depending on the concentration of the element determined in the sample. The slide valve is now pushed into the other position. This directs the carrier gas to the gas sampling line and flushes the gas volume trapped in the sampling loop into the pyrolysis tube. Typical carrier gas flow rates are 40 ml/min for oxidative analysis and 10 ml/min for reductive analysis. When the titration of the sample has been completed, i.e. the recorder pen has returned to the base line, the slide valve is pushed back to the initial position and the sampling device is ready for the next sample.

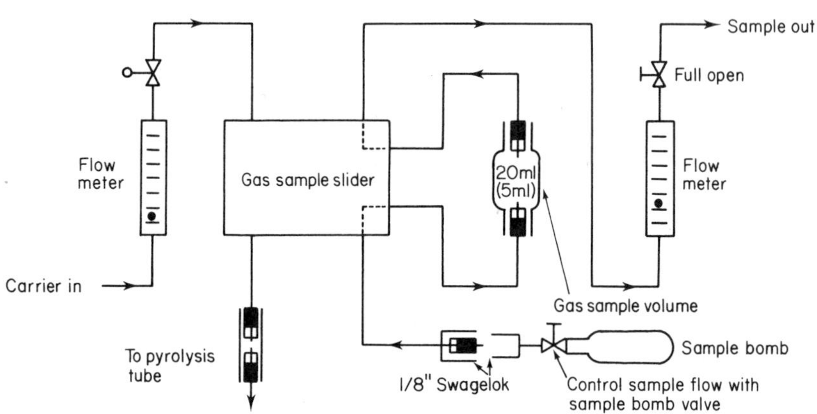

FIG. 21. Scheme of sample inlet for gases (Dohrmann).[79]

The analysis of LPG (liquefied petroleum gas) presents particular difficulties. First, it is necessary to sample the liquid phase present in the pressurized sample container, because the gaseous phase has a different composition than the liquid phase, depending on the relative vapour pressures of the sample components. Furthermore, withdrawal of the vapour phase will disturb the equilibrium between the phases and alter the composition of both phases. Secondly, the sudden drop in pressure when opening the valve on the inverted sample container produces a cooling effect which in turn causes fractionation of the emerging vapours, which again alters the original sample composition. Thus, sampling of LPG for analysis is a major problem which is best solved by measurement of the sample volume in the liquid form, followed by expansion of the sample into the pyrolysis tube. This can be achieved with the system shown in Fig. 21 by eliminating the sample loop and using the bore of the slide valve to determine the liquid LPG volume. Volumes less than 10 μl arc usually adequate. The inverted pressurized sample container is connected to the slide valve, the control valve A is closed and the valve on the sample container opened. The sample flow rate is now adjusted by slowly opening valve A and the sample introduced into the pyrolysis tube by operating the slide valve as described above.

G. Other accessories

A heating tape is recommended to prevent the condensation of water, present in the sample or formed by the pyrolysis processes in the capillary cell inlet, and absorption of polar gases such as HCl or SO_2 in the water. A 30–60 W heating tape and an adequate power supply are required to maintain the temperature of the inlet capillary at about 90°C.

It is advisable to use a dummy cell in the form of a constant voltage source when checking the performance characteristics of the electronic sections of the titration system.[(48)]

An electrode plating unit should also be available when using the microcoulometric system for the determination of chlorine or nitrogen. In the first case the sensor electrode and generator anode are made of silver-plated platinum; in the second case platinum electrodes are used with platinum black deposited on them. Prolonged use and repeated cleaning of the electrodes makes it necessary to replace them from time to time. The plating baths and procedures are described later.

Coulometer testing and electrode plating units are available commercially.*

The gas supply lines must be equipped with two-stage pressure regulators and flowmeters. The latter are usually built into the furnace unit.

IV. DETERMINATION OF SULPHUR (OXIDATIVE METHOD)

The method is intended primarily for the determination of sulphur in the range 1 to 200 ppm in liquids that can be introduced into the pyrolysis tube with a syringe and completely volatilized at 550°C. The method can be applied to concentrations up to 10000 ppm, but at concentrations above 200 ppm it is advisable to check the linearity of the coulometer response or to dilute the sample with an adequate solvent. At concentra-

* A-100 Coulometer Testing Unit (range 0–1 V) and E-250 Electrode Plating Unit, Dohrmann Envirotech, Mountain View, Cal.

tions below 1 ppm, special measures and precautions are necessary to ensure adequate sensitivity and precision (see Section IV. G). As described, the method is intended for samples of predominantly hydrocarbon character, containing sulphur in the form of hydrogen sulphide, thiols, organic sulphides and disulphides, or thiophene derivatives. The method may also be applied to non-hydrocarbon matrices and other sulphur species but it has to be kept in mind that the presence of heteroelements such as nitrogen, chlorine and oxygen may influence the combustion equilibria and interfere with the reactions in the coulometric cell. In such cases the applicability of the method should be checked on model substances.

Gases may be analyzed by the method as described, but LPG (liquefied petroleum gas) requires a special sample inlet (Section III. F). The method can be extended to heavy liquids and solids with the use of sample boats (Section III. F).

A. Outline of procedure

A suitable amount of sample is burned with oxygen in the combustion tube. The combustion products are swept into the titration cell where sulphur dioxide formed by the combustion of sulphur compounds is titrated with electrogenerated iodine

$$I_3^- + SO_2 + H_2O \rightarrow SO_3 + 3I^- + 2H^+ \tag{16}$$

The titrant ion I_3^- consumed in this reaction is regenerated according to the reaction

$$3I^- \rightarrow I_3^- + 2e^- \tag{17}$$

and the current, recorded as a function of time is a measure of the sulphur present in the sample. Carbon dioxide and water formed in the combustion of hydrocarbons have no effect on the cell reactions, nitrogen and halogen compounds and heavy metals if present in the sample may interfere in the determination.

B. Apparatus and materials

The apparatus required for the determination consists of a combustion unit, the microcoulometer and the titration cell, as shown by the block scheme in Fig. 10. The combustion is carried out in an empty pyrolysis tube, such as that shown in Fig. 11. The titration cell used for this determination is a T-300-P iodine cell (Fig. 18). Auxiliary equipment includes a recorder and integrator, a magnetic stirrer, a heating tape to prevent condensation of water in the capillary tube connecting the combustion tube exit to the titration cell, and a grounded cabinet to protect the cell from electrostatic pick-up effects.

All chemicals should be reagent grade and the water used in preparing the cell electrolyte should be demineralized or distilled.

1. *Cell electrolyte*

The electrolyte (Fig. 18) is prepared by dissolving 0·5 g of potassium iodide and 0·6 of sodium azide in 1000 ml of deionized water and adding 5 ml of glacial acetic acid. The electrolyte should be stored in a brown bottle or in a dark place. Shelf life varies from a few weeks to a few months, depending on reagent purity and other factors.

2. *Gases*

The reactant gas is laboratory grade oxygen, and the carrier gas is preferably laboratory grade argon or helium. Nitrogen may be used instead but should be tested for the presence of impurities which react with iodine.

3. *Sulphur standards*

Solutions of GPR grade thiophene (b.p. 84·2°C) or di-n-butyl sulphide (b.p. 182°C) in sulphur-free isooctane or toluene have been found suitable. Cyclohexane sulphide and benzyl thiophene have also been recommended.[80,81] The best way is to prepare a stock solution of ca. 1000 μg/ml S, which is then further diluted as required to match the sulphur concentration of the samples. It is a recommendable precaution to use, wherever possible, standards that correspond not only to the sample concentration but also to the boiling range and sulphur compound type and to inject the same volume of standard and sample for the determination.

Sulphur standards prepared from a pure compound remain stable for several months. Standards diluted to very low concentrations (below 10 μg/ml) must be prepared at more frequent intervals.

C. Preparation of apparatus and operating conditions

1. *Preparation of apparatus*

The titration cell is prepared for operation as described in Appendix I.1.

Place the cell centrally on the magnetic stirrer and connect the capillary gas inlet to the combustion tube. Attach the gas supply lines to the combustion tube and make the electrical connections between the microcoulometer, the cell and the recorder and integrator, according to the manufacturer's instructions and the block scheme in Fig. 10. Connect the power supplies to the furnace, the magnetic stirrer, the capillary heater and the electronic equipment.

Adjust the gas flow rates, the furnace temperatures and the microcoulometer bias and gain to the operating conditions chosen.

2. *Operating conditions*

Typical operating conditions are given in Table 3.

Table 3. Operating conditions for the determination of sulphur by the oxidative method

Gas flow rates	
Oxygen	120–160 cm^3/min
Carrier gas	30–50 cm^3min
Furnace temperatures	
Inlet	450–550°C
Centre	850–950°C
Outlet	800–850°C
Sample size	
Concentration 1–20 μg/ml S	10–3 μl
Concentration 20 μg/ml S	3–1 μl
Microcoulometer	
Bias voltage	140–160 mV
Amplifier gain	100–400

Note: A 15-mm quartz wool plug should be inserted in the outlet section of the combustion tube.

The flow rates of the oxygen and the carrier gas are chosen so that a ratio of oxygen to carrier gas of 3 : 1 to 4:1 is obtained. A high partial pressure of oxygen does not favour the formation of SO_2 in the combustion but is necessary to achieve a satisfactory combustion of the organic matrix. An oxygen-to-carrier gas ratio of 1:4 has also been recommended,[22,80,81] but an excess of oxygen in the combustion gases is now generally preferred.[82]

A temperature of 450–550°C in the inlet section of the combustion tube has been chosen to guarantee complete vaporization of samples boiling not higher than 400°C. Temperatures much higher than 550°C may induce cracking of the sample matrix with the formation of carbon deposits in the inlet section. A high temperature in the middle section of the tube, where the actual combustion of the sample takes place, has been selected to give a favourable conversion to SO_2 (about 80% at 900°C), but not high enough to seriously shorten the lifetime of the quartz combustion tube. A temperature of 800–850°C in the outlet section of the tube maintains the SO_2/SO_3 equilibrium without affecting the cell temperature.

A sample size of 1–3 μl is satisfactory for most samples with a sulphur content above 20 μg/ml. For sulphur contents below 20 μg/ml, the sample size must be increased with decreasing sulphur concentration: 10 μl are adequate at the 1 μg/ml level and as much as 40 μl may be required for sulphur contents below this level.[82] If the standby technique[83, 75] is used, 10 μl of sample is sufficient for sulphur contents from 0–4 μg/ml (see Section IV. G). The maximum amount of sample which the combustion tube can accept when injected rapidly is about 2 μl. Larger samples must be injected at a carefully controlled rate not exceeding 0·3 μl/s.

Scaringelli *et al.*[84] have shown, by the use of SO_2 permeation tubes, that the current efficiency of the iodine cell is 100%. Their findings were confirmed by Drushel.[61] Taking $0{\cdot}5 \times 10^{-8}$ M as the optimum concentration of the titrant ion (I_3^-) in the cell electrolyte, a value of 160 mV can be calculated for the bias voltage from electrochemical data.[48] However, the bias voltage may be varied between 160 and 140 HV without affecting the response and the stability of the cell. At lower bias voltages the response of the cell becomes sluggish, and at voltages above 160 mV the cell becomes unstable, i.e. it loses its capacity of maintaining a constant titrant concentration. The amplifier gain and the range in ohms are selected to give a satisfactory peak shape and size at a minimum of background noise (see below). The optimum value varies with sulphur concentration, sample size, amplifier characteristics, etc.

3. *Procedure*

A determination is carried out as described in Appendix I.2. The injection of a sample causes a peak to appear on the recorder chart which may have one of the shapes shown in Fig. 22. The operating conditions are varied in small steps as described in the Appendix until the correct shape B is obtained. Typical peaks obtained by repeated injection of the same sample are shown in Fig. 23. Broad, flat-topped peaks are obtained at high sulphur concentrations when the regeneration rate of the cell is exceeded.[20, 23]

4. *Calculations*

The sulphur content of the sample is calculated from the equation:

$$\text{Sulphur } (\mu\text{g/ml}) = \frac{C \times A}{S} \tag{18}$$

where C = concentration of standard, in μg/ml
A = area of sample peak, in integrator units/range in ohm/μl
S = area of standard peak, in integrator units/range in ohm/μl

To obtain the sulphur concentration in parts per million (ppm) mass, divide the concentration in μg/ml with the density of the sample.

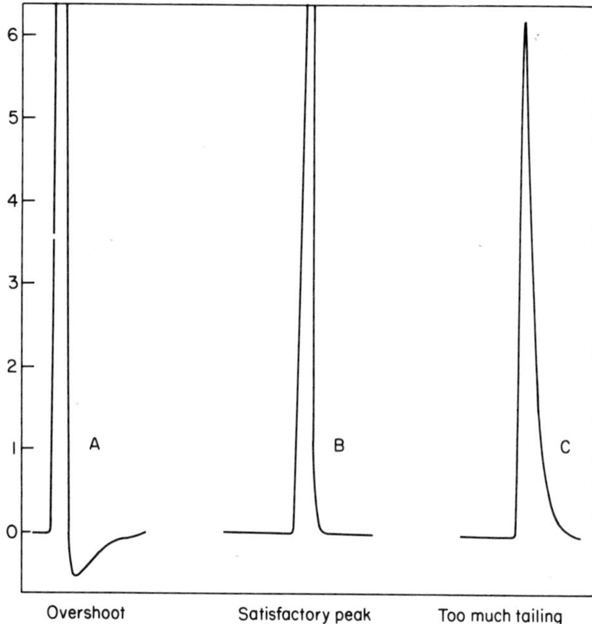

FIG. 22. Microcoulometric peak shapes: (A) Overshoot, (B) Satisfactory peak, (C) Too much tailing.

If the microcoulometer output is displayed directly in nanograms (this is the case on the Dohrmann C-300 microcoulometer), the recovery factor F of a standard solution is determined:

$$F = \frac{\text{ng shown by integrator}}{\text{ng injected}}$$

and the sulphur content of the sample is calculated from the equation

$$\text{S (ng)} = \text{ng on integrator}/F \tag{19}$$

The recovery factor F multiplied by 100 gives the total recovery of the system in percent.

The sulphur content of a sample represented by a peak displayed on a recorder chart can be calculated from electrochemical data. From Faraday's law (Eqn (1)) it follows that

$$Q = It = \frac{V}{R} LA \times 10^{-3} \tag{20}$$

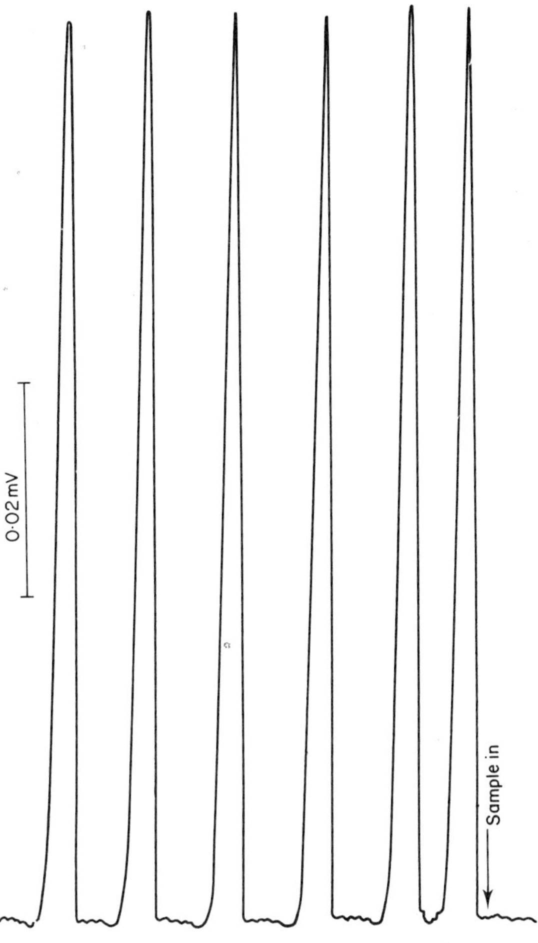

FIG. 23. Typical response peaks for sulphur by the oxidative method. Each peak represents the injection of 1 μl of naphtha containing 96 μg/ml S. (Bias 155 mV, range 20 phms, gain ca. 400).

where Q is the quantity of electricity used in the titration in coulombs, I is the current in amperes, t is the time in seconds, V is the recorder sensitivity in mV cm^{-1}, R is the resistance in ohms, L is the chart speed in s. cm^{-1} and A is the peak area in cm^2.

Substitution of this equation in Eqn (2) gives

$$W\,(\mathrm{ng}) = \frac{VLAM \times 10^6}{R \times 96\,500 \times nZ} \qquad (21)$$

where W is the weight of the substance titrated in nanograms, A is the peak area in arbitrary integrator units, M is the molecular weight (in this case of sulphur), n is the number of electrons involved per mole of electrode reaction, 96 500 is the value of the Faraday in coulombs and Z is the count rate in arbitrary units/cm^2, which has to be determined separately for a given recorder–integrator combination.

For the titration of SO_2 with iodine, taking into account the recovery factor F for this determination, Eqn (21) takes the form

$$\mathrm{S\ (ng)} = \frac{VLA}{RZF} \times 165{\cdot}8 \tag{22}$$

where F is the recovery factor, determined separately on a standard solution:

$$F = \frac{\text{ng S calculated}}{\text{ng S injected}} = \text{ca. } 0{\cdot}80$$

D. Precision and accuracy

Precision data for the oxidative sulphur method are shown in Table 4.

Table 4. Precision of oxidative microcoulometric method for sulphur

Sample	No. of determinations	Sulphur (ppm) mean value	range (ppm)	SD (± ppm)	RSD[a] (± %)
Solutions of thiophene in toluene					
	9	158·0	8	3·5	2·2
	5	92·6	3	1·3	1·5
	6	72·3	4	1·6	2·3
	6	47·7	1	0·5	1·1
	6	19·5	1	0·6	2·8
	7	7·3	0·5	0·3	3·7
Naphtha					
	20	581·8	37	13·0	2·2
	19	292·0	15	4·2	1·4
	12	219·5	9	3·8	1·7
	16	111·4	8	2·0	1·8
	18	53·5	6	1·9	3·5
Kerosene					
	6	659·8	25	8·3	1·3
	10	496·5	29	9·6	1·9
Diesel oil					
	10	0·566[b] wt%	0·062	0·020	3·5
	10	0·28[b] wt%	0·04	0·011	3·9

[a] RSD = relative standard deviation.
[b] After dilution with toluene.

At concentrations above 100 ppm the relative standard deviation (RSD) is ±1 to 3% and slightly higher if the sample had to be diluted prior to the determination.

Below 100 ppm the standard deviation increases with decreasing sulphur concentration (82):

Sulphur (ppm)	SD (±ppm)	RSD (±%)
100·0	3·20	3·2
10·0	0·50	5·0
1·0	0·10	10·0
0·1	0·05	50·0

Typical results obtained by the method on solutions of thiophene in toluene and petroleum fractions are given in Table 4.

The accuracy of the method can be assessed by comparing the results obtained by the microcoulometric method with results obtained by other methods. Typical data are shown in Table 5.

Table 5. Comparison of sulphur determined by microcoulometry with results obtained by other methods

Sample	Method	ppm S	ppm S by microcoulometry
Thiophene in iso-octane 1	Wickbold[a]	0·5	0·57
2	Wickbold	45·5	50·0
Heptanes	Wickbold	11·1	11·7
Toluene	Wickbold	1·5	1·33
Naphtha 1	Wickbold	30·0	32·3
2	Wickbold	91·0	93·0
3	Lamp[b]	300	292
4	Lamp	340	325
5	XRF[c]	600	582
Jet Fuel 1	Lamp	181	162
2	Lamp	1300	1400
Kerosine	XRF	192	186
Diesel oil	Wickbold	320	349
Gas oil 1	XFR	0·23 wt %	0·231 wt %
2	XFR	0·80 wt %	0·800 wt %
Base oil 1	XRF	0·25 wt %	0·249 wt %
2	XRF	0·29 wt %	0·297 wt %

[a] Oxy-Hydrogen Combustion, ASTM Method D2785—IP Method 243.
[b] Lamp Combustion, ASTM Method D 1266—IP Method 107.
[c] X-Ray Fluorescence, ASTM Method D 2622.

E. Interferences

Halogen and nitrogen compounds present in the sample are likely to interfere in the titration of sulphur dioxide with iodine. The interference is caused by the oxidation of nitrogen compounds to nitrogen oxides and of halogen compounds to halogen (Section II. C); they liberate iodine from the potassium iodide present in the electrolyte, thus increasing the iodine concentration in the electrolyte, and produce a "negative" peak.

If the sample contains no sulphur or only in low concentrations, the recorder pen drops below the baseline; at higher sulphur concentrations the results will be too low, the difference being a function of the concentration of the interfering element in the sample. This effect is illustrated by the data in Table 6, obtained with a solution of 103 μg/ml thiophene in hexane, to which varying amounts of *n*-butylamine, pyridine, 1,2-dichloroethane or 1,2-dibromoethane were added.

The data show that nitrogen begins to interfere at concentrations near the 1 g/litre (0·1%) level, chlorine at 0·5 g/litre (0·05%) and bromine already at the low ppm level.

As suggested by Bremanis and co-workers,[86] the interference caused by nitrogen and chlorine can be eliminated by the addition of sodium azide to the cell electrolyte (see Table 6). This is based on the fact that sodium azide reacts rapidly with chlorine and oxides of nitrogen to give chloride and molecular nitrogen, respectively, while with iodine, on the other hand, it reacts very slowly unless catalysed by bivalent sulphur compounds:

$$SO_2 + H_2O \longrightarrow 2H^+ + SO_3^- \quad (23)$$

$$SO_3^- + I_2 \longrightarrow I^- + ISO_3^- \quad (24)$$

$$ISO_3^- + 2N_3^- \longrightarrow I^- + SO_3^- + 3N_2 \quad (25)$$

$$I_2 + 2N_3^- \longrightarrow 2I^- + 3N_2 \quad (26)$$

Bromine has a much stronger effect on sulphur determination than chlorine because a much smaller fraction is converted to HBr in the combustion (less than 50%, see Section II. C). This interference cannot be overcome by the addition of sodium azide to the electrolyte, because its reaction with bromine is too slow.

Metals if present in the sample will form oxides in the combustion. These may combine with the sulphur trioxide and shift the SO_2/SO_3 equilibrium, thus causing low sulphur results. The addition of sodium azide to the electrolyte has no effect on this interference.

For these reasons motor gasolines cannot be analyzed by the oxidative

Table 6. Effect of nitrogen and halogen compounds on the determination of sulphur by the oxidative method

Interfering element, (g/litre)	Sulphur determined (μg/ml) Electrolyte I (without Na azide)	Electrolyte II (with Na azide added)
N (as *n*-Butylamine)		
0	101	—
0·07	100	—
0·71	101	—
1·41	97	—
2·78	95	—
6·78	81	100
N (as Pyridine)		
0	102	—
0·09	104	—
0·87	101	—
3·41	102	—
8·29	85	102
Cl (as 1,2-Dichloroethane)		
0	103	104
0·49	100	—
0·97	99	106
1·94	97	104
4·86	90	106
9·71	80	103
Br (as 1,2-Dibromoethane)		
0	103	104
0·034	99	100
0·068	95	94
0·14	87	90
0·27	69	70
0·34	56	61
0·68	7·6	16

[a] Electrolyte I: 0·05% KI, 0·5% acetic acid.
[b] Electrolyte II: same, plus 0·06% sodium azide.

method, since they contain lead anti-knock compounds (usually tetraethyl lead or tetramethyl lead) and lead scavengers (1,2-dichloroethane alone or in mixture with 1,2-dibromoethane). The latter compounds are added to the lead alkyls to form in the combustion the volatile lead chloride and lead bromide instead of the nonvolatile lead sulphate which tends to deposit on spark plugs and thus affects engine performance.

F. Applications

The method has been applied to a wide range of products of predominantly hydrocarbon character, ranging from gases to liquids and solids. These include natural gas,[18, 19, 85] light and middle petroleum distillates and hydrocarbon solvents,[18–20, 22, 28, 52, 54, 85] and white oils with boiling points from 315 to 565°C.[18, 19, 85] Killer and Underhill[20] determined nonvolatile sulphur in liquefied petroleum gases (most of which is elemental sulphur) by passing the liquid LPG from an inverted sample cylinder into sulphur-free toluene, allowing the volatile components to weather off, and determining the sulphur content of the remaining toluene solution. Dixon (52) modified the sample inlet of the combustion tube and used it to determine the sulphur content of fuel oils and fractions obtained from preparative thin-layer chromatography by direct combustion in oxygen, using an inlet temperature of 800°C. Bandi *et al.*[87] determined the sulphur content of iron and steel down to 1 ppm by combustion in a zirconia tube heated to 1500°C and microcoulometric titration of the sulphur dioxide evolved with iodine.

De Groot *et al.*[49] used a microcoulometric system of their own design (see Section III. B) to determine sulphur in pesticide solutions at concentrations ranging from 5 to 100 μg/ml by combustion and titration with electrogenerated iodine. The electrolyte was prepared by 5:1 dilution of a stock solution made up of 0·2 g of sodium iodide, 20 g of sodium bromide, 5 g of glacial acetic acid and 6 g of sodium azide. The sulphur recovery (at combustion and outlet zone temperatures of 950°C) was about 76% and independent of sulphur compound type but varied with the amount of sulphur titrated (from 78·8% at 5 ng/μl to 74% at 100 ng/μl).

Cedergren[50] determined sulphur in liquid hydrocarbons under conditions selected so that the recovery as SO_2 was close to 100% (low partial pressure of oxygen, temperature of 1000°C in the equilibrium zone). The relative standard deviation was less than 1% in the range 2–1000 μg/ml sulphur, with thiophene in cyclohexane as test substance.

Underhill estimated the distribution of sulphur compound types in kerosines and jet fuels by the method of Ball,[88] based on the sequential treatment of the sample with chemicals which selectively remove certain sulphur compound types. Between treatments the total sulphur content of the sample is determined by the microcoulometric method and the quantity of the sulphur type in question obtained from the difference between the sulphur content before and after treatment. In this relatively simple form, the method provides orientative data on the distribution of sulphur compound types in petroleum distillates. The following chemical treatment is used:

Treatment	*Removes*
(1) Water wash	water-soluble S compounds
(2) Shaking with mercury	elemental S
(3) Refluxing with acetic acid and zinc	RSSR, by reduction to RSH
(4) Shaking with mercury(I) nitrate crystals	RSR

Thiols are determined separately by potentiometry. Subsequent treatment with mercury(II) nitrate determines aromatic sulphides and thiophenes as a group. Unfortunately, a large fraction of the sulphur present in the sample remains unclassified and this fraction increases with the increasing boiling point of the sample.

Constant-current coulometry instead of null-balance microcoulometry has also been used to determine sulphur in organic media by combustion and titration of the sulphur dioxide evolved, e.g. by Carter[(89)] to determine sulphur in petroleum oils in the range 0·01 to 1%, by Hoshino[(90)] and Glass and Moore,[(91)] who applied high-frequency combustion and coulometric titration to determine 0·05–4·8% sulphur in petroleum oils, and by Fraisse and Raveau.[(92)]

G. Determination of sulphur at concentrations below 1 ppm (Standby Technique)

As described, the microcoulometric method offers a quick and simple way of determining sulphur in volatile products down to levels around 1 ppm. However, at levels below 1 ppm the conditions required for this determination are very difficult to maintain and the precision of the results is affected by excessive background noise and various phenomena connected with the combustion of the sample. The writer studied the factors affecting the determination and found that rigid control of sample combustion was required to avoid spurious pressure, temperature and/or adsorption-desorption effects that were superimposed on the actual SO_2 titration peaks.[(83)] At these very low sulphur levels peak shape and area, base line stability and background noise are strongly influenced by combustion tube design, combustion temperature, titration cell temperature, and the condition of the combustion tube and titration cell. To obtain peaks of adequate size, large samples (up to 40 μl) must be combusted and the combustion must be carried out slowly, because of the limited capacity of the combustion tube. An extended combustion period, in turn, leads to superposition of combustion and titration which results in low and broad peaks that are difficult to integrate. However, the superposition of the two processes can be avoided by interrupting the feedback from the amplifier to the generating electrodes during combustion of the sample. As a result of this, titration of the SO_2 formed in the combustion is delayed until interfering phenomena in the cell have ceased and equilibrium is re-established. Sharp peaks are obtained when the regeneration circuit is switched on again, at operating conditions that are easily maintained and reproduced. A blank peak is also obtained, mainly due to impurities in the gases, which is a linear function of time (at given furnace and cell temperatures). This makes it necessary to carry out the analysis in a controlled time interval. Typical peaks obtained with this technique are shown in Fig. 24.

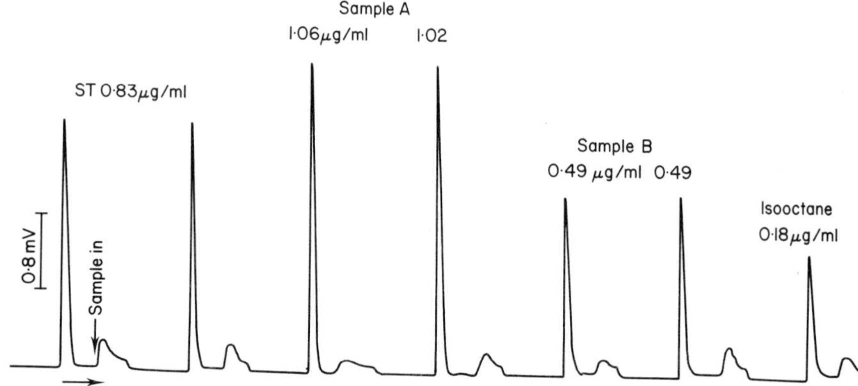

FIG. 24. Typical sulphur peaks obtained with the standby technique.[75] (Sample size 10 μl, Bias 150 mV, range 100 ohms, gain ca. 100). The small broad peaks are needle peaks.

The equipment required for the determination of sulphur by the standby technique is essentially the same as that used for the standard method. The following points are important when working at sulphur levels near or below 1 ppm:

1. A microcoulometer is required which allows a temporary switch-off of the regeneration circuit. The Dohrmann C-200 and LKB 16 300 microcoulometers are suitable, but the Dohrmann C-300 model does not possess this facility.

2. A slow and controlled combustion of the samples is essential to obtain repeatable results. This is easily achieved by the use of a cranking device (see Section III. F).

3. The insertion of the empty syringe needle (with the sample withdrawn into the syringe barrel) into the hot inlet section of the combustion tube pro-

duces a "needle peak" which is negligible at higher sulphur levels but becomes quite prominent at levels below 5 ppm (see Fig. 24). This peak is partly due to sample vapours in the syringe needle. It is also related to the furnace inlet temperature and the adsorption of oxygen on the hot needle, withdrawn from the furnace after an injection. The lower the inlet temperature, the lower the needle peak. Hence the lowest inlet temperature should be chosen that gives complete vaporization of the sample.

4. The capillary inlet to the titration cell must be heated in low-ppm work to prevent condensation of water formed in the combustion of the sample and absorption of SO_2 in this water, which would lead to low-frequency baseline fluctuations and peaks of irregular shape (Section III. G).

5. A grounded shield for the cell is required to eliminate electrostatic pick-up which increases the background noise. The shield must be designed so that it does not prevent the dissipation of heat from the cell and the stirrer. A Faraday cage made of perforated sheet metal or wire gauze or a small grounded open-ended metal box are satisfactory; they provide sufficient air circulation to maintain the cell at a constant temperature near room temperature. A cabinet which completely encloses the cell and the stirrer is not recommended.

6. To prevent the generation of heat which may raise the cell temperature (see Section III. C), it is recommended to use a magnetic stirrer with a separate speed regulator located outside the shield. An increase in electrolyte temperature results in higher blank values.

7. The same considerations with regard to combustion tube design, com-

Table 7. Operating conditions for the determination of sulphur by the Standby Technique

Gas flow rates	
Oxygen	100–150 cm^3/min
Carrier gas	30–50 cm^3/min
Furnace temperatures	
Inlet	450°C
Centre	850°C
Outlet	800°C
Sample size	
Concentration 0–4 μg/ml S	10 μl
Microcoulometer	
Bias voltage	145–160 mV
Amplifier gain	120

Note: A 15 mm quartz wool plug should be inserted in the outlet section of the combustion tube.

bustion temperatures and gas flow rates apply as for the standard method.

The operating conditions recommended for the determination of sulphur by the standby technique are given in Table 7.

A determination is carried out as described in Appendix I.3. A linear plot of peak area versus sulphur concentration is obtained as shown in Fig. 25.

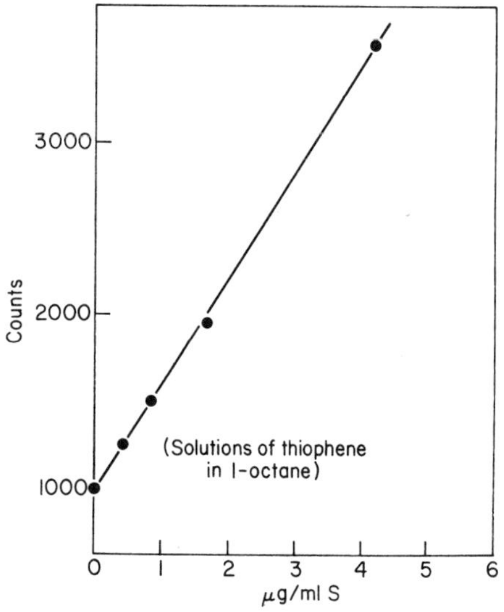

FIG. 25. Determination of sulphur by the standby technique: response versus sulphur concentration.[75]

The precision of the results obtained by the standby technique was found to be very satisfactory. Two parallel determinations differed only rarely by more than 0·1 μg/ml. Similarly, in the analysis of blends with a known sulphur content, made up of thiophene or di-*n*-butyl sulphide in heptane, iso-octane or toluene, the average of two determinations was, as a rule, within 0·1 μg/ml of the blended value (Δ). Typical results are presented in Table 8.

An average standard deviation of ±0·04 μg/ml was calculated from the results available. The method has been applied successfully to hydrofined and reformed naphthas with sulphur concentrations of 0–4 μg/ml. It may also be applied to the microcoulometric determination of chlorine or nitrogen at trace levels.

Table 8. Precision of sulphur determinations by the Standby Technique (blends of thiophene or di-*n*-Butyl sulphide in iso-octane)[75]

Sample	Sulphur, μg/ml			Δ[a]
	Taken	Found	Mean	
1	0·07	0·06, 0·09	0·06	0·01
2	0·22	0·15, 0·25, 0·18	0·19	0·03
3	0·50	0·43, 0·46	0·45	0·05
4	0·83	0·89, 0·82	0·86	0·03
		0·74, 0·78	0·76	0·07
		0·87, 0·87	0·87	0·04
5	1·03	1·08, 1·06	1·07	0·04
		1·01, 1·09	1·05	0·02
6	1·68	1·60, 1·68	1·64	0·04
7	2·00	2·00, 1·94	1·97	0·03
8	2·28	2·21, 2·27	2·24	0·04
		2·27, 2·30	2·29	0·01

[a] Difference between value blended and mean value found.

V. DETERMINATION OF SULPHUR (REDUCTIVE METHOD)

The method is intended primarily for the determination of total sulphur in the range 1 to 200 parts per million in liquids that can be introduced into the pyrolysis tube with a syringe and completely volatilized at 550°C. The method can be applied to concentrations up to 10 000 ppm,[93] but at concentrations above 200 ppm the sample should be diluted with an adequate solvent as titration time becomes excessive. Gases may be analyzed by the method as described, but LPG (liquefied petroleum gas) requires a special sample inlet (Section III. F). Viscous liquids and solids can be analyzed when introduced into the pyrolysis tube with a boat, preferably as solutions (Section III. F). The method is not applicable where the nitrogen content of the sample is greater than 10 times the sulphur content or if the chlorine content exceeds 50%.

A. Outline of procedure

The sample is injected into the pyrolysis tube where it vaporizes, mixes with a flow of humidified hydrogen and passes through the reaction zone packed with a platinum catalyst kept at 1150°C. The sample undergoes destructive pyrolysis, whereby the sulphur is converted to hydrogen sulphide, which is titrated with electrogenerated silver ion

$$2Ag^+ + S^{2-} \longrightarrow Ag_2S \qquad (27)$$

The titrant Ag^+ consumed in this reaction is regenerated according to the reaction

$$Ag^0 \longrightarrow Ag^+ + e^- \tag{28}$$

and the current, recorded as a function of time, is a measure of the amount of sulphur present in the sample.

B. Apparatus and materials

The apparatus required for the determination consists of a pyrolysis unit, the microcoulometer and the titration cell, as shown by the block scheme in Fig. 10. The pyrolysis is carried out in a tube such as that shown in Fig. 12. packed with the hydrogenation catalyst. The titration cell for this determination is a T-400-S silver cell (Fig. 18). Auxiliary equipment includes a recorder and integrator, a magnetic stirrer, a heating tape to prevent condensations in the capillary cell inlet, a grounded cabinet to protect the cell from electrostatic pick-up and a gas humidifier. The use of a gas manifold such as the one shown in Fig. 26 is recommended, to reduce the hazard involved in working with hydrogen at elevated temperatures.

All chemicals should be reagent grade and the water used in preparing the electrolyte should be demineralized or distilled.

1. *Catalyst*

The central section of the pyrolysis tube is packed with a catalyst consisting of 10% platinum on alundum (10–20 mesh). The catalyst is prepared as described in Appendix II.1.[22]

2. *Cell electrolyte*

The cell electrolyte (see Fig. 18) is made up of 13·6 g sodium acetate ($CH_3COONa \cdot 3H_2O$) and 17 ml conc ammonia in one litre of deionized water. It is best stored in a Pyrex bottle. Plastic squeeze bottles should not be used as they often release plasticizers into the electrolyte and contaminate the electrode surface.

3. *Gases*

High-purity hydrogen must be used as the reactant gas, preferably supplied in cylinders. Alternatively, it could be produced by a high-pressure electrolytic generator and/or purified with a palladium diffusion purifier.

A gas blend of 10% NH_3 in nitrogen is required to maintain a given ammonia concentration in the cell electrolyte, and laboratory-grade oxygen and argon or helium must be at hand for periodic catalyst reconditioning.

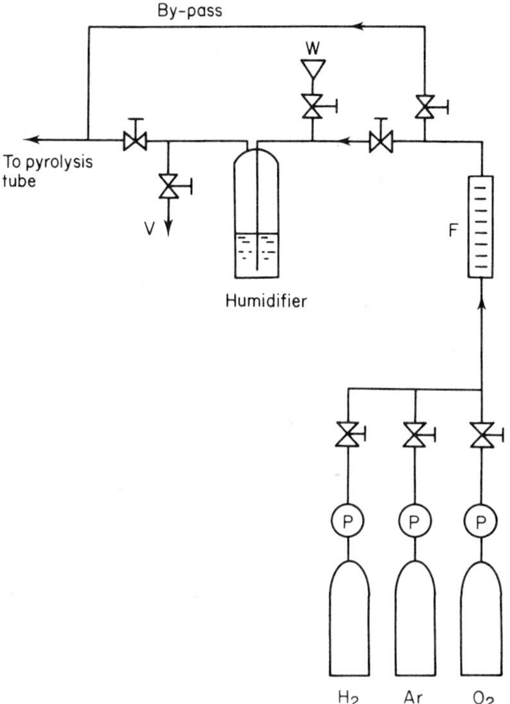

FIG. 26. Gas manifold: F, flow meter; W, sidearm for water addition: V, vacuum (as required); P, two-stage pressure regulators; H, Drechsel-type humidifier with coarse frit.

4. *Sulphur standards*

The same sulphur standards can be used for the reductive method as for the oxidative method (Section IV. C).

C. Preparation of apparatus and operating conditions

1. *Preparation of apparatus*

The pyrolysis tube is prepared for operation as described in Appendix II.2 To recondition a catalyst that has already been used proceed as described in Appendix II.3.

The titration cell is prepared for operation as described in Appendix II.4.

Place the cell centrally on the magnetic stirrer and connect the main capillary gas inlet to the pyrolysis tube. Attach the gas supply lines to the pyrolysis tube and the cell. Make the electrical connections between the microcoulometer, the cell and the recorder (and integrator) according to the manufacturer's instructions and the block scheme in Fig. 10. Connect the power supplies to the furnace, the magnetic stirrer, the capillary heater and the electronic equipment.

Adjust the gas flow rates, the furnace temperatures and the microcoulometer bias and gain to the operating conditions chosen.

2. *Silver plating of electrodes*

The performance of the silver-plated sensor and generator anode may deteriorate in prolonged use or due to damage of the electrode surface. They can be replated using a variable d.c. source with a current output of 0–200 mA. The plating procedure is given in Appendix II.5.

3. *Operating conditions*

Typical operating conditions are given in Table 9.

Table 9. Operating conditions for the determination of sulphur by the reductive method

Gas flow rates	
Hydrogen (humidified)	300 cm^3/min
10% NH_3 in nitrogen	40 cm^3/min
Furnace temperature	
Inlet	550–700°C
Centre	1150°
Outlet	750°
Sample size	1–8 μl
Microcoulometer	
Bias	100–110 mV
Gain	100–400

In this method hydrogen serves both as the carrier and the reactant gas. Prior to entering the pyrolysis tube the gas is passed through a Drechsel-type humidifier, because the presence of water in the pyrolysis zone aids removal of carbon and reduces its deposition on the catalyst surface by "the water–gas" reaction (Section II. D):

$$H_2O\,(g) + C \xrightarrow{>600^\circ C} CO + H_2 \qquad (29)$$

Alternatively, water can be injected into the hydrogen stream at a preset rate by means of a syringe pump.

Farley and Winkler[95] investigated the conditions for the reductive pyrolysis of sulphur compounds and found that complete and reproducible recovery of the sulphur compounds as hydrogen sulphide is obtained only if very little deposition of uncombusted hydrocarbon is allowed in the pyrolysis tube. The variables controlling this deposition are the hydrogen flow rate, the degree of humidification and the dimensions of the combustion tube.

An inlet temperature of 550°C is sufficient to volatilize low-boiling liquids. However, when boats are used to introduce heavy liquids or solids, the inlet temperature must be raised to 800°C. The very high temperature of 1150° in the catalyst section shortens drastically the life of the quartz pyrolysis tube which devitrifies rapidly at this temperature (within 2–3 weeks). However, the life of the tube can be extended to three months or more by lowering the temperature of this section to 900° when the instrument is not in use.[22] The relatively low outlet temperature of 750° was selected because dissociation of H_2S was found to occur at higher temperatures,[22] resulting in reduced H_2S recovery.

A sample size of 1–8 μl is adequate for most determinations. The injection rate must not exceed 0·2 μg/ml. When more than 200 ng of sulphur are injected, the sample should be diluted with a suitable solvent to a concentration near 100 ppm.

The bias potential chosen (110 mV) and the presence of ammonium ion in the electrolyte make it possible to operate the cell at a very low silver ion concentration ($pAg^+ = 12$). When the concentration of ammonia is maintained at 0·3 M, most of the silver is present in the form of the complex ion $Ag(NH_3)_2^+$. The ratio $Ag^+/Ag(NH_3)_2^+$ is equal to 10^{-6}, so that only one in 10^6 Ag^+ ions is available for reaction with other ions. This makes the cell selective for sulphide ion in the presence of chloride, since at a silver ion concentration of 10^{-12} M the chloride concentration must reach 100 M before precipitation can occur. Cyanide ions interfere in the determination.

A determination is carried out as described in Appendix II.6.

4. Calculations

The sulphur content of the sample is calculated from Eqns (18) or (19) or directly from electrochemical data, using Eqn (21).

For the titration of hydrogen sulphide with silver ions this equation takes the form

$$S(ng) = \frac{VLA}{RZF} \times 165{\cdot}8 \tag{29}$$

where V is the recorder sensitivity in mV cm^{-1}, L is the chart speed in s. cm^{-1}, A is the peak area in integrator units, R is the range in ohms, Z is the integrator count rate in units/cm^2, and F is the recovery factor, which has to be determined separately on a standard solution (see Section IV.D). For this determination $F \approx 1{\cdot}00$.

D. Precision and accuracy

The precision data in Table 10 are considered as typical for the method.

Table 10. Precision of reductive microcoulometric method for sulphur.[93]

Sulphur (ppm)	SD (±ppm)	RSD[a] (± %)
>500	15·0	3·0
100	3·5	3·5
10	0·8	8·0
1	0·5	50·0

[a] RSD = Relative standard deviation.

The accuracy of the method can be assessed from measurements made on synthetic blends and by comparing the results obtained by the microcoulometric method and other methods. Typical data are shown in Table 11.

E. Interferences

Nitrogen compounds are likely to interfere with the determination, due to the formation of HCN which is titrated with the silver ions under the cell conditions selected (see Section II. D). The amount of HCN formed is proportional to the carbon-to-nitrogen ratio in the sample; in a hydrocarbon medium about 4% of the nitrogen present is converted to HCN which will be recorded as roughly the same apparent amount of sulphur (1 ppm N as HCN corresponds to 1·14 ppm S).[22] The rest of the nitrogen is converted to ammonia.

F. Applications

The method has been tested on various sulphur compounds[22, 24] and has been applied to petroleum products ranging from light distillates to residual fuel oils.[22]

G. Comparison between oxidative and reductive sulphur methods

A comparison between the two methods for the microcoulometric determination of sulphur must take into account the following advantages and disadvantages of the two methods.

Table 11. Comparison of sulphur determined by microcoulometry with results obtained by other methods.[22]

Sample		Method	ppm S	ppm S by microcoulometry
Di-*n*-butyl disulphide in kerosine			3[a]	2·4
			12[a]	10·0
Thiophen in cyclohexane			3[a]	2·8
			12[a]	11·8
Thiophen in toluene			3[a]	2·9
			12	2·9
Light and middle distillates	1	XRF[b]	300	320
	2	XRF	400	510
	3	XRF	600	570
	4	XRF	0·12 wt %	0·124 wt %
	5	XRF	0·40	0·424
	6	XRF	0·76	0·805
	7	XRF	1·15	1·16
	8	XRF	1·30	1·34
	9	XRF	1·43	1·48
	10	XRF	1·51	1·42
Residual fuel oils	1	Quartz tube[c]	1·35	1·23[d]
	2	Quartz tube	1·40	1·35[d]
	3	Quartz tube	1·41	1·47[d]
	4	Quartz tube	1·42	1·36[d]

[a] Blended values.
[b] X-Ray fluorescence, ASTM Method D 2622.
[c] Quartz tube combustion, ASTM method D 1551—IP method 63.
[d] Sample introduced with boat.

1. The oxidative method is generally easier to perform and the equipment requires less attention (e.g. no catalyst to regenerate).

2. The use of hydrogen at high temperatures in the reductive method represents a safety hazard.

3. The very high pyrolysis temperature of 1150°C required for the reductive method accelerates devitrification of the quartz tube and thus makes frequent tube replacement necessary.

4. Nitrogen and halogen compounds interfere with the oxidative sulphur determination. Addition of sodium azide to the electrolyte overcomes this interference, except in the case of bromine. Heavy metals may also have an effect on the results. On the other hand, halogen compounds have no effect on the reductive sulphur determination, but nitrogen compounds interfere. This interference can be suppressed by humidification of the hydrogen but not completely overcome.

5. The precision and accuracy of the two methods are similar.

6. The recovery of sulphur in the oxidative method is not quantitative; only about 80% of the sulphur is converted to SO_2 and thus available for titration. The recovery of sulphur as H_2S in the reductive method approaches the stoichiometric value. Data obtained by Wallace, Joyce and co-workers[(22)] on solutions of pure sulphur compounds illustrate this point (Table 12).

Table 12. Sulphur recovery obtained by the oxidative and reductive methods (Data of Wallace, Joyce *et al.* [(22)])

Sulphur compound	b.p. (°C)	Oxidative (syringe)	Reductive (syringe)	Reductive (boat)
(Solutions in kerosine, 10–100 ppm)				
Thiophen	84·2	84	100	100
Di-*n*-butyl sulphide	187	87	100	102
Di-*n*-butyl disulphide	226	84	94	100
Di-tert-octyl sulphide	—	79	88	103
Dodecanethiol	142^{15}	86	102	102
Elemental sulphur	445	73	86	100
(Solutions in butanol, 10–100 ppm)				
4-Toluenesulphonic acid	140^{20}	75	91	100
Dibutyl sulphone	decomp	79	84	101
Diethyl sulphate	208	68	80	100
Dimethylsulphoxide	189	93	100	100

Note: Inlet temperatures for syringe injections 550°C, for boat injections 800°C.

The recoveries of both methods obtained when the samples are injected by syringe vary appreciably. The low recoveries are attributed to hold-up of the samples in the syringe rather than to inefficient pyrolysis. This is supported by the fact that stoichiometric recoveries were obtained with the reductive method when the samples were introduced into the pyrolysis tube by means of a sample boat, using an inlet temperature of 800°C.

Braier *et al.*[(24)] carried out a very thorough comparison of the two sulphur methods which included a statistical evaluation of the results obtained. The tests were carried out on toluene solutions of the following sulphur compounds, containing 1, 5, 10, 25, 100 ppm sulphur:

	b.p. (°C)
Dibenzothiophene	332
1-Hexanethiol	151
Diphenyl sulphide	296

Dipropyl disulphide	194
Sulpholane (Tetrahydrothiophene 1,1-dioxide)	285
Thiophene	84·2

A qualitative inspection of the data showed that more consistent replicate results were obtained with the reductive than with the oxidative method except at the 1 ppm level. The data of Braier *et al.* are rearranged in Table 13

Table 13. Sulphur recovery obtained by the oxidative and reductive methods (Data of Braier *et al.*[24])

Compound	Sulphur, ppm						Average recovery, %
	1	5	10	25	50	100	
Dibenzothiophene							
Oxidative	110	74	80	80	84	80	84·7
Reductive	70	96	101	92	96	100	92·5
1-Hexanethiol							
Oxidative	53	34	49	80	80	79	62·5
Reductive	64	94	98	96	94	98	90·7
Diphenyl sulphide							
Oxidative	110	74	80	84	84	86	86·3
Reductive	80	96	100	88	98	101	93·8
Dipropyl disulphide							
Oxidative	33	42	41	88	78	82	60·7
Reductive	74	82	91	92	98	100	89·5
Sulpholane							
Oxidative	59	82	81	84	82	85	78·8
Reductive	74	92	94	92	98	100	91·7
Thiophene							
Oxidative	57	74	77	84	78	82	75·3
Reductive	68	94	95	96	102	98	92·2
Average recovery %							
Oxidative	70·3	63·3	68·0	83·3	81·0	82·3	
Reductive	71·7	92·3	96·5	92·7	97·7	99·5	

to show the sulphur recoveries obtained with the two methods. Each figure represents an average of three determinations. The data show that the average recoveries for each sulphur compound obtained by the oxidative method vary significantly, mainly because of the very low recoveries obtained for 1-hexanethiol (62·5%) and dipropyl disulphide (60·7%). Neglecting these two compounds, the recoveries of the others are in the range 75·3–86·3%, as could be expected for a conversion to SO_2 of about 80% at the operating conditions used. The low recoveries of the two compounds mentioned are, however, limited to the concentration range 1–10 ppm; at concentrations 25–100 ppm they are near to 80%, indicating that the low recoveries may be due to methodical rather than structural causes. This is particularly true for the recoveries at the 1-ppm level which, in the case of the oxidative method, vary over a wide range (33–110%) and in the case of the reductive method are consistently low (64–80%).

The recoveries for each compound obtained by the reductive method vary from 89·5 to 93·8%. Neglecting the low results obtained at the 1-ppm level, the recoveries are much closer to the expected 100 per cent.

In conclusion it can be said that the reductive method seems to be less affected by various factors such as sulphur recovery, matrix and structural effects than the oxidative method, thus giving a somewhat better precision. However, the oxidative method is more widely used for practical reasons, such as ease of operation and maintenance, etc.

VI. DETERMINATION OF CHLORINE

The method is intended for the determination of total chlorine concentrations in the range 1 to 500 ppm in liquids which can be completely volatilized at 550°C. At concentrations above 200 ppm it is advisable to check the linearity of the coulometer response or to dilute the sample with an adequate solvent. Heavy liquids and solids can also be analyzed, preferably as solutions, by using a boat inlet to introduce the samples into the combustion tube (see Section III.F). Bromine or iodine as well as high concentrations of nitrogen interfere in the determination.

A. Outline of the procedure

A suitable amount of sample is burned with oxygen in the combustion tube. The combustion products are swept into the titration cell where the hydrogen chloride formed by the oxidative pyrolysis of chlorine compounds is titrated with electrogenerated silver ions

$$Ag^+ + Cl^- \rightarrow AgCl \tag{30}$$

The titrant ion Ag^+, consumed in this reaction, is regenerated according to the reaction

$$Ag^0 \rightarrow Ag^+ + e^- \tag{31}$$

and the current, recorded as a function of time, is a measure of the chlorine present in the sample.

B. Apparatus and materials

The apparatus required for the determination is essentially the same as that used for the determination of sulphur by the oxidative method (see Section IV.B and the block scheme in Fig. 10). The titration cell used for this determination is a T-300-S silver cell (Fig. 18). Auxiliary equipment includes a recorder and integrator, a magnetic stirrer, a grounded cabinet to protect the cell from electrostatic pick-up and direct light, and a heating tape to prevent condensations in the capillary tube which connects the combustion tube with the cell. Heating of the cell inlet is essential in this determination (see below).

All chemicals should be reagent grade and the water used in preparing the cell electrolyte should be demineralized or distilled.

1. Cell electrolyte

The electrolyte is 70 vol% aqueous acetic acid and can be stored indefinitely in a Pyrex bottle. The electrolyte should not be kept in a plastic squeeze bottle since plasticizer is leached from the polymer and can contaminate the electrode surfaces.

2. Gases

The reactant gas is laboratory grade oxygen; the carrier gas is preferably laboratory grade argon or helium. Nitrogen may be used instead, but should be tested for the presence of impurities which react with silver ions.

3. Chlorine standards

Solutions of GPR grade chlorobenzene (b.p. 132°C) or 1,2-dichloroethane (b.p. 83·6°C) in iso-octane or toluene are suitable. The best way is to prepare a stock solution of ca. 1000 μg/ml Cl, which is then further diluted as required to match the chlorine concentration of the samples. Standards remain stable for several months, except those diluted to very low concentrations (below 10 μg/ml), which must be prepared at more frequent intervals.

C. Preparation of apparatus and operating conditions

1. *Preparation of apparatus*

The titration cell is prepared for operation as described in Appendix III.1.

Place the cell centrally on the magnetic stirrer and connect the capillary gas inlet to the combustion tube. Attach the gas supply lines to the combustion tube and make the electrical connections between the microcoulometer, the cell, and the recorder and integrator according to the manufacturer's instructions and the block scheme in Fig. 10. Connect the power supplies to the furnace, the magnetic stirrer, the capillary tube heater and the electronic equipment.

Adjust the gas flow rates, the furnace temperatures and the microcoulometer bias and gain to the operating conditions selected.

2. *Operating conditions*

Typical operating conditions are given in Table 14.

Table 14. Operating conditions for the determination of chlorine

Gas flow rates	
Oxygen	120–160 cm^3/min
Carrier gas	30–50 cm^3/min
Furnace temperatures	
Inlet	550°C
Centre	850–950°C
Outlet	800–850°C
Sample size	
Concentration 1–20 μg/ml Cl	40–5 μl
Concentration above 20 μg/ml Cl	5–1 μl
Microcoulometer	
Bias voltage	240–265 mV
Gain	1000–2400

Note: A 15-mm quartz wool plug should be inserted in the outlet section of the combustion tube.

The flow rates of the oxygen and carrier gas are chosen so that a ratio of oxygen to carrier gas of 3:1 to 4:1 is obtained. An inlet temperature of 550°C is selected to achieve complete volatilization of samples without cracking of the sample matrix. This may occur at higher temperatures, leading to the formation of carbon deposits in the inlet section where there is no oxygen. At the above oxygen-to-carrier gas ratios and flow rates, the combustion and outlet temperatures selected give an almost complete conversion of organic chlorine to HCl (about 98%).

A sample size of 1–5 μl is satisfactory for most samples with chlorine contents above 20 μg/ml. At chlorine contents near 2 μg/ml as much as 40 μl of sample may be required for a determination. Samples larger than 2 μl must be injected slowly, at a carefully controlled rate not exceeding 0·3 μl/min. The amount of chlorine injected should not exceed 500 ng, because above this level titration time becomes excessive. Instead, the sample should be diluted to a concentration near 100 ppm.

A bias voltage of 250 mV can be calculated from electrochemical and solubility data for an optimum silver ion concentration of 10^{-7} mol/litre. However, the cell is stable and can be operated over the range 240–280 mV.[48]

High amplifier gains are required for this determination, therefore, all factors contributing to background noise must be carefully controlled.

Krijgsman *et al.*[68] used an 80% acetic acid electrolyte to which they added 1% of sodium perchlorate. This increased the impedance of the system and thus reduced the sensitivity of the cell towards electrical disturbances.

An important factor in the determination of chlorine is the tendency of HCl formed in the combustion to be retained by combustion water, condensing in the capillary inlet tube of the cell. This portion of the HCl is then released and titrated as a separate peak when the water evaporates. The effect is avoided by heating of the cell inlet to about 90°C. Typical chlorine peaks obtained with and without cell inlet heating are shown in Fig. 27.

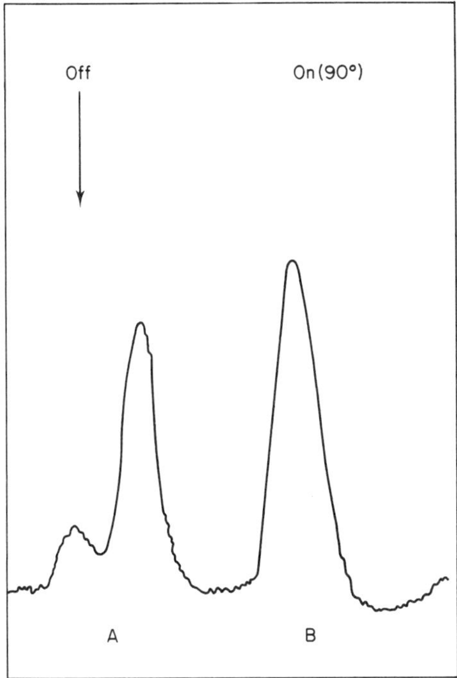

FIG. 27. Effect of cell inlet heating on shape of chlorine peak: (A) With no heating, (B) with inlet tube heated to 90°C.

Humidification of the gases has been proposed[11, 55] to increase the conversion of organic chlorine compounds to HCl. However, the HCl recovery is already high (98%) so that a further increase has little effect on the overall performance of the system.

The response of silver-plated electrodes becomes sluggish after prolonged use due to silver halide adhering to the electrode surface. The electrodes can be reactivated by a short treatment with concentrated ammonia and nitric acid and thorough rinsing with deionized water.[68] If this treatment fails, the electrodes must be replated (see Appendix II.5).

The silver cell must be protected from daylight, which decomposes silver chloride to form chloride, which is then again titrated by the coulometer.[68]

A determination is carried out as described in Appendix III.2. Typical chlorine peaks are shown in Fig. 28.

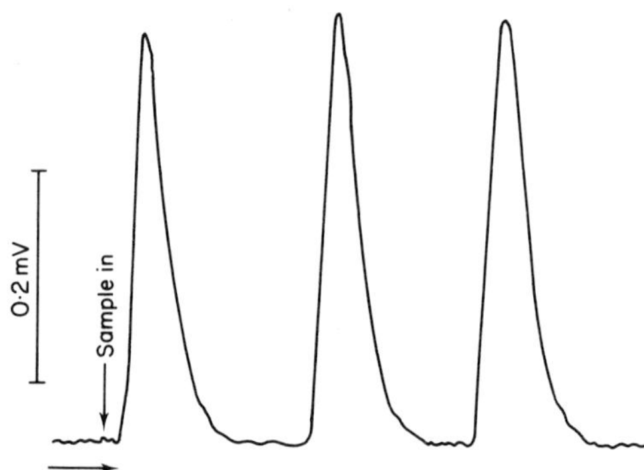

FIG. 28. Typical response peaks for chlorine. Each peak represents the injection of 5 µl of naphtha containing 11.8 µg/ml Cl. (Bias 268 mV, range 100 ohms, gain ca. 2000).

3. *Calculations*

Calculate the chlorine content of the sample from Eqns (18) or (19). Alternatively, the chlorine content of the sample may be derived from electrochemical data by using Eqn (21) on p. 201, which for the titration of hydrogen chloride with silver ion takes the form

$$\mathrm{Cl(ng)} = \frac{VLA}{RZF} \times 367{\cdot}5 \qquad (32)$$

where V is the recorder sensitivity in mV cm^{-1}, L is the chart speed in s. cm^{-1}, A is the peak area in arbitrary integrator units, R is the resistance in ohms, Z is the count rate in arbitrary units/cm^2 for the recorder-integrator combination used for the analysis, and F is the chlorine recovery factor, determined separately on a standard solution ($F \approx 1{\cdot}00$).

D. Precision and accuracy

The precision data shown in Table 15 have been quoted for the method.

Table 15. Precision of the microcoulometric method for chlorine[(96)]

Chlorine (ppm)	SD (± ppm)	RSD[a] (± %)
500	15·0	3·0
100	3·2	3·2
10	0·50	5·0
1	0·10	10·0

[a] RSD = Relative standard deviation.

The accuracy of the method can be estimated by comparing the results obtained by the microcoulometric and other methods. Typical data are shown in Table 16.

Table 16. Comparison of chlorine determined by microcoulometry with results obtained by other methods

Sample	Method	ppm Cl	ppm Cl by microcoulometry
Naphtha 1	Na diphenyl[a]	1·2	1·0
2	Na diphenyl	1·4	2·5
3	Na diphenyl	3·0	3·0
4	Na diphenyl	7·0	7·4
5	Na diphenyl	8·8	11·2
Data from Drushel[(18)]			
Cl compound in diethyl ether	Lamp[b]	26	32
Cl compound in ethanol	Lamp	10	11
Ethanol	Lamp	0·5	0·2
Isopropanol (91 %)	Lamp	0·5	3·8
Data from Lädrach *et al.*[(59)]			
Distillate (350–480°C) 1	Wickbold[c]	5·4	4·9[d]
2	Wickbold	25·8	25·5[d]
3	Wickbold	3·8	2·0[d]
Benzene	Wickbold	1·0	0·8[d]
White oil	Wickbold	1·5	1·3[e]
Gasoline (phosphorus-containing)	Flask[f]	1800	2190[g]

[a] B. Pecherer, C. M. Gambrill and G. W. Wilcox, *Anal. Chem.* **22** (1950), 311.
[b] Lamp combustion, ASTM Method D 1266—IP Method 107
[c] Oxy-hydrogen combustion, ASTM method D 2785—IP method 243
[d] Average of three determinations
[e] Average of five determinations
[f] Flask combustion, IP method 244
[g] Average of six determinations

E. Interferences

Nitrogen compounds when present in concentrations larger than $6{\cdot}5 \times 10^4$ times the chlorine concentration[96] are likely to interfere in the titration of chlorine with silver ions. The interference is caused by nitrogen oxides formed in the combustion, resulting in a negative peak. Krijgsman *et al.*[68] eliminated this effect by the addition of a few milligrams of sulphamic acid to the electrolyte in the cell. These authors also tested the effect of some metalloids on the determination and found that sulphur, phosphorus and arsenic had no effect, even if present in large quantities.

Bromine and iodine give a positive interference, as they are partially converted in the combustion to the corresponding hydrogen halides which react with the silver ions.

Lädrach *et al.*[59] pointed out that water present in the sample or formed in the combustion of the organic matrix dilutes the cell electrolyte and thus alters the silver potential. Additional silver ions must be generated to restore equilibrium. They found that the 4–6 mg of water formed in the combustion of 5 µl of hydrocarbon sample produced an apparent chlorine peak of about 3 ppm. To overcome this difficulty the water was trapped between the combustion tube and the cell in an absorber filled with 80% sulphuric acid. The system was tested on synthetic blends of myricyl chloride in kerosine with chlorine contents ranging from 1 to 43 ppm and chlorine recoveries of 90–100% were obtained.

Cedergren and Johansson[74] studied this dilution effect in more detail and suggested two alternative ways to compensate for the end-point displacement: one is to determine a solvent blank and to subtract it from the sample count; the other is to set the bias at the potential at which the titration curves for the diluted and the undiluted electrolytes intersect.

F. Applications

The method has been applied to solvents,[18, 19, 59, 69] light and middle distillates,[25, 59] organochlorine pesticides[26] and other organic substances.[57, 69] Solutions of inorganic chlorides in kerosine–isopropanol blends were also analyzed:[59] chlorides which volatilize or decompose at 400°, such as ammonium chloride, iron(III) chloride and palladium chloride gave complete recoveries, but the more stable chlorides, e.g. those of alkali metals, magnesium and tin(II), were not recovered.

Samples with chlorine contents below 1 ppm were analyzed by trapping the hydrochloric acid formed in the combustion of large samples (e.g. 100 µl) in a cold trap at −80°C and liberating it by subsequent rapid heating of the trap to 300°C.[59] Killer[133] discussed the microcoulometric determination of chlorine (and sulphur) in organic materials.

Solomon and Uthe[134] described a method for the determination of chlorine in organic materials by combustion and coulometry. A combustion apparatus capable of burning both solid and liquid samples is used where the samples are vaporized in helium and burned in humid oxygen at 820–900°C. The HCl formed in the combustion enters a condenser–absorber unit, made of a length of 2-mm i.d. Teflon tubing wound around a core, and the condensate collected is flushed with a few ml of the electrolyte (acetic acid–nitric acid) straight into a titration vial. The chloride content of the solution is then determined coulometrically, using a constant-current coulometer (Aminco clinical chloride titrator).

McFee and Bechtold[135] combined the Mast continuous coulometric analyzer[123] with a pyrolysis unit to determine a variety of halogenated hydrocarbons with toxic properties in air. The lower limits of detection ranged from 2 parts per billion for methyl iodide to 900 ppb for tetrachlorodifluoroethane. The system was found to be insensitive to benzyl chloride and chlorinated biphenyls.

VII. DETERMINATION OF NITROGEN

The method is primarily intended for the determination of total nitrogen, bound in organic or inorganic form in the range 1 to 200 parts per million in liquids that can be completely volatilized at 550°C. It can be extended to concentrations below 1 ppm as well as to concentrations up to 5000 ppm.[97] In the latter case it is advisable to check the linearity of coulometer response and/or to dilute the sample with an adequate solvent. Gases may be analyzed by the method as described, higher boiling liquids and solids require a boat inlet (Section III.F). The method is not applicable to samples containing more than 5% sulphur or 20% halogen.

A. Outline of procedure

The sample is introduced into a stream of humidified hydrogen and brought into contact with a nickel catalyst maintained at 800°C. Reductive pyrolysis converts the bound nitrogen to ammonia which is titrated with electrogenerated hydrogen ions:

$$NH_3 + H^+ \rightarrow NH_4^+ \qquad (33)$$

The titrant ion (H^+) consumed in this reaction is regenerated according to the reaction

$$\tfrac{1}{2}H_2 \rightarrow H^+ + e^- \qquad (34)$$

and the current, recorded as a function of time is a measure of the amount of nitrogen present in the sample.

Acid gases formed in the pyrolysis are absorbed in an alkaline scrubber before they reach the titration cell.

The reaction at the generator cathode produces hydroxyl ions:

$$H_2O + e^- \rightarrow OH^- + \tfrac{1}{2}H_2 \qquad (35)$$

The cathode must therefore be isolated from the cell cavity by means of a diffusion barrier.

B. Apparatus and materials

The apparatus required for the determination consists of a pyrolysis unit, the microcoulometer and the titration cell, as shown by the block scheme in Fig. 10. The pyrolysis is carried out in a tube such as that shown in Fig. 12, packed with the hydrogenation catalyst and equipped with a scrubber tube (Fig. 13). The titration cell for this determination is a T-400-H hydrogen cell (Fig. 18). Auxiliary equipment includes a recorder and integrator, a magnetic stirrer, a heating tape to prevent condensation in the capillary cell inlet, a grounded shield to protect the cell from electrostatic pick-up and a gas humidifier.

A gas supply manifold such as that shown in Fig. 26 is also recommended.

All chemicals should be reagent grade and the water used in preparing the cell electrolyte should be demineralized or distilled.

1. *Catalyst*

The central section of the pyrolysis tube is packed with nickel granules 20–50 mesh or foil clippings 0·1 mm $\times$ 10 mm^2. The nickel metal should contain at least 99·6% Ni.

2. *Cell electrolyte*

Dissolve 10 g sodium sulphate (anhydrous) in one litre of water and store in a Pyrex bottle.

3. *Gases*

High-purity hydrogen must be used as the reactant gas, preferably supplied in cylinders. Titratable impurities should not exceed 5 ppm. Alternatively, hydrogen produced by a high-pressure electrolytic generator may be used, purified with a palladium diffusion purifier.

Laboratory grade oxygen and argon or helium must be at hand for periodic catalyst reconditioning.

4. *Nitrogen standards*

Pyridine or other heterocyclic nitrogen compounds in iso-octane or toluene are suitable. The best way is to prepare a stock solution of 200–1000 ppm nitrogen, which is further diluted to match the nitrogen concentration of the samples.

C. Preparation of apparatus and operating conditions

1. *Preparation of apparatus*

The pyrolysis tube is prepared for operation as described in Appendix IV.1. The reconditioning of a catalyst that has become contaminated through extended or improper use is described in Appendix IV.2.

The acid gas scrubber is prepared for use as described in Appendix IV.3. Preparation of the titration cell for operation is described in Appendix IV.4.

The lead plating procedure for the reference electrode is given in Appendix IV.5.

The procedure to be followed for the deposition of carbon black on the sensor electrode is given in Appendix IV.6. If in operation the cell gives a noisy baseline, the electrode is plated for additional 5 seconds. In general, a thicker platinum black coating gives a quieter operation but more overshoot.

Place the cell centrally on the magnetic stirrer and connect the main capillary gas inlet to the pyrolysis tube. Attach the hydrogen supply line to the pyrolysis tube. Make the electrical connections between the microcoulometer, the cell and the recorder and integrator according to the manufacturer's instructions and the block scheme in Fig. 10. Connect the power supplies to the furnace, the magnetic stirrer, the capillary heater and the electronic equipment.

Adjust the gas flow rates, the furnace temperatures and the microcoulometer bias and gain to the operating conditions chosen.

2. *Operating conditions*

Typical operating conditions are given in Table 17.

In this method hydrogen serves both as the carrier and the reactant gas. Hydrogen flow rate has an effect on peak shape; a higher flow rate gives sharper peaks with less tailing. Moistening of the hydrogen also reduces tailing of peaks, by suppressing coke formation on the catalyst and ammonia adsorption on the coke. With dry hydrogen, ammonia adsorption increases as coke accumulates and results in low ammonia recoveries. Too much moisture causes a loss in catalyst activity and gives again low ammonia recoveries. Correct humidification of the hydrogen stream is best achieved by splitting the hydrogen supply into two streams, one of which is passed through a scrubber filled with water. A gas manifold like the one shown in Fig. 26 is

Table 17. Operating conditions for the determination of nitrogen

Gas flow rates	
Humidified hydrogen	20–50 cm^3/min
Total humidified + dry hydrogen	400 cm^3/min
Furnace temperatures	
Inlet	550–700°C
Centre	750–800°C
Outlet	300°C
Sample size	
Concentration > 10 μg/ml N	3–15 μl
10–1 μg/ml	10–20 μl
< 1 μg/ml	20–30 μl
Microcoulometer	
Bias	100–120 mV
Gain	350–450 mV

recommended for this purpose. Such a manifold also reduces the hazard involved in working with hydrogen at high temperatures.

Several catalysts were tested by Fabbro *et al.*[98] for hydrogenation capacity, ammonia conversion and peak shape. The granulated nickel catalyst gave the best overall results with ammonia recoveries of 95–100%. Aavik *et al.*[99] obtained full ammonia conversion with a nickel spiral at 850–900°C in dry hydrogen and on a nickel–MgO catalyst at 440°C with water-saturated hydrogen. The former catalyst was preferred since the Ni-MgO catalyst affected the cell equilibrium.

An inlet temperature of 550°C is sufficient to volatilize low-boiling liquids. Higher temperatures must be used for heavier liquids and with the boat inlet (800–1000°C). Care must be taken to avoid cracking of the samples in the inlet section, because this leads to losses of nitrogen due to inclusion in the coke deposits formed, as pointed out by Spies and Harris.[100]

The catalyst operates best at ca. 800°C. At lower temperatures, coking occurs on the catalyst; at higher temperatures nitrogen is retained, leading to peak broadening.

The outlet section of the pyrolysis tube is kept at 300°C to suit the scrubber tube packed with sodium hydroxide on alundum. The strong alkali attacks the inner surface of the scrubber tube, causing it to craze and eventually break. Periodical inspection and testing for mechanical strength is therefore required. Other absorbents such as CaO,[29,64] MgO[30,99] and potassium carbonate[18,101] have also been used, but NaOH is now generally preferred.

An injection of 3–12 μl will be adequate for most samples. Larger volumes must be injected when determining low concentrations of nitrogen, but low nitrogen recoveries can be expected because of the time taken to inject the sample. Contrary to the expected behaviour, higher ammonia conversions are obtained with rapid than with slow injection. This may be due to coke formation occurring before the pyrolyzed sample reaches the catalyst.[18] Injection rates of up to 1 μl/s have been suggested.[97] Sample introduction with a boat may overcome this problem in many cases. This way of introducing the sample into the pyrolysis tube has also the advantage that no "needle peak" is produced, as the sample is injected into the boat in the cool zone, i.e. the syringe needle does not get hot. This is of particular importance in the analysis of samples with very low nitrogen contents, where the needle peak can be much larger than the sample peak. Samples containing more than 200 ppm nitrogen should be diluted to about 50 ppm with a suitable solvent to avoid low nitrogen recoveries.

The titration cell works most satisfactorily, i.e. the peaks are the narrowest without significant overshoot, when the hydrogen ion concentration in the electrolyte is kept at pH 6. If the hydrogen ion concentration is reduced ($pH > 6$), the sensitivity increases but the stability of the cell decreases and the peaks overshoot. If the hydrogen ion concentration increases ($pH < 5{\cdot}5$), the system becomes insensitive, resulting in low and broad peaks. The response of the cell is also a function of sodium sulphate concentration; concentrations down to 0·4 g/litre Na_2SO_4 may be used, but the microcoulometer gain has to be increased, leading to an increase in background noise. With an 1% sodium sulphate solution the gain can be reduced, thus also reducing the noise level.

A determination is carried out as described in Appendix IV.7. Typical nitrogen peaks are shown in Fig. 29.

3. *Calculations*

Calculate the nitrogen content of the sample from Eqns (18) or (19). Alternatively, the nitrogen content of the sample may be derived from electrochemical data by using Eqn (21) (Section IV.C) which for the titration of ammonia with hydrogen ion takes the form:

$$\mathrm{N(ng)} = \frac{VLA}{RZF} \times 145{\cdot}1 \tag{36}$$

where V is the recorder sensitivity in mV cm^{-1}, L is the chart speed in s. cm^{-1}, A is the peak area in integrator units, R is the range in ohms, Z is the integrator count rate in units/cm^2, and F is the recovery factor which has to be determined separately on a standard solution (Section IV.D). For this determination $F = 0{\cdot}90$–$1{\cdot}00$.

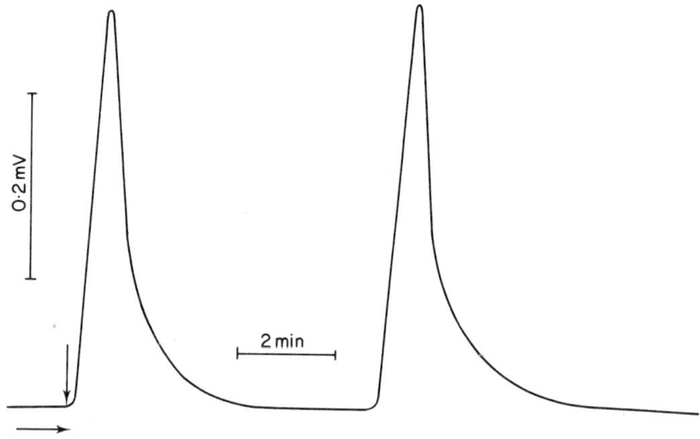

FIG. 29. Typical response peaks for nitrogen. Each peak represents the injection of 5 μl of a solution of pyridine in toluene, containing 35 μg/ml N. (Bias 110 mV, range 20 ohms, gain ca. 400).

D. Precision and accuracy

The following precision has been quoted for the method (Table 18).

Table 18. Precision of the microcoulometric method for nitrogen.[97]

Nitrogen (ppm)	SD (± ppm)	RSD[a] (± %)
500·0	—	3·0
100·0	3·20	3·2
10·0	0·50	5·0
1·0	0·10	10·0
0·5	0·07	14·0
0·1	0·05	50·0

[a] RSD = Relative Standard Deviation.

Nitrogen recoveries obtained on synthetic blends of nitrogen compounds were near to 100% as can be seen from the data in Table 19.

Low recoveries were obtained with azobenzene, presumably due to the fact that azo compounds tend to rupture the N–N bond with the formation of molecular nitrogen instead of reduction to ammonia. This view is supported by the observation that higher recoveries are obtained at low inlet temperatures (and high hydrogen flow rates); Fabbro *et al.*[98] report a recovery of 95% for azobenzene at an inlet temperature of 400°C.

Table 19. Nitrogen recovery obtained on synthetic blends of nitrogen compounds

Compound	N Recovered as ammonia, %	
	(a)	(b)
Pyridine	98	94
Octylamine	99	96
Indole	101	106
Quinoline	102	106
Aniline	100	100
Nitrobenzene	—	96
Azobenzene	40	80
Heptyl cyanide	—	100

(a) Solutions in iso-octane 20 and 200 ppm N, data of Fabbro *et al.*[98]
(b) Solutions in kerosine 50 ppm N, data of Gouverneur and van de Craats.[30]

When analyzing solutions of octylamine in kerosine, Gouverneur and van de Craats[30] found that nitrogen recoveries started to decrease at nitrogen concentrations above 250 ppm. This observation supports the suggestion to reduce the nitrogen content of samples to below 200 ppm by dilution with a solvent prior to analysis.

Satisfactory recoveries were also obtained with organic nitrogen compounds and inorganic salts in aqueous solution, as shown by the data in Table 20.

Table 20. Nitrogen recovery obtained on aqueous solutions of organic nitrogen compounds and inorganic salts

Compounds	N Recovered as ammonia, %	
	(a)	(b)
Pyridine	98	—
Diethamolamine	100	94–102
Aminotriazole	90	—
Ammonium sulphate	—	97–103
Potassium nitrate	100	100
Calcium nitrate	100	—
Nickel nitrate	99	—

(a) 20 and 200 ppm N, data of Fabbro *et al.*[98]
(b) 10 ppm N, data of Albert *et al.*[29]

When samples with oxygen-containing materials such as aldehydes, alcohols or water are analyzed, the cell response is first acidic, followed by the basic response of the ammonia. This acid cell response is due to CO_2 produced in the pyrolysis[(28)] and must be eliminated by retaining it on the alkaline packing in the scrubber tube.

Good agreement is generally achieved between results obtained by the microcoulometric and the Kjeldahl methods for nitrogen. Typical data obtained with the two methods on a range of products are presented in Table 21.

Table 21. Comparison of nitrogen determined by the microcoulometric and Kjeldahl methods

Sample	Nitrogen, ppm Kjeldahl	Nitrogen, ppm Microcoulometer mean	Deviation from mean
Data from Drushel[(18)]			
Hydrogen treated products			
(boiling range 325–460) 1	106	105	±2
2	139	142	1·5
3	146	145	1
4	147	141	1
5	138	126	1
6	126	123	2·5
7	99	96	0·5
Heavy coker naphtha	345	252	2
Treated lube stocks 1	1288	1265	8
2	745	752	15
Data from Fabbro *et al.*[(98)]			
Raw kerosine	1·5[a]	0·93	±0·02
Premium gasoline 1	4·0	2·9	0·05
2	14·0	10·1	0·01
Light gas oil	28·9	21·2	0·1
Motor Oil	18·0	21·1	0·3
No. 2 fuel oil 1	47·8	51·0	1·5
2	87·5	92·8	0·5
Commercial additive	63·5	63·9	1·5
Data from Albert *et al.*[(29)]			
Raw sewage	28	25	±2
Primary effluent 1	48	51	1
2	12·7	11·4	0·5
Secondary effluent 1	3·9	4·2	0·3
2	5·2	6·4	0·7

[a] Results obtained with a Kjeldahl method using a modified sulphuric acid extraction, proposed by O. I. Milner, R. J. Zahner, L. S. Hepner, and W. A. Cowell, *Anal. Chem.* **30** (1958), 1528.

E. Interferences

The nitrogen determination is based on the titration of NH_3 with hydrogen ion. Any strong acid reaching the cell will therefore cause a negative interference if it is not efficiently absorbed by the scrubber tube. The system as described is not suitable for samples containing more than 20% of halogen or 5% of sulphur.

F. Determination of nitrogen in materials boiling above 550°C

The method described above is limited to the determination of nitrogen in liquids that can be completely volatilized at 550°C. The major difficulties encountered with higher boiling samples are incomplete volatilization or coking in the inlet section of the pyrolysis tube. Both causes lead to low nitrogen recoveries. To overcome this difficulty, Drushel[(61)] modified the pyrolysis tube so as to permit direct injection of high boiling liquids through a narrow capillary onto the catalyst with a minimum of coke formation. Using this modified tube and an inlet temperature of 825°C, nitrogen was determined in residua containing asphaltenes with an average recovery of 95%, as compared with the Kjeldahl method. In order to reduce plugging of the needle and to obtain complete sample injection, the sample in the syringe was backed with a small volume of toluene.

Rhodes *et al.*[(101)] used a 1·2 m long quartz tube coil heated to 700°C as the inlet section for their pyrolysis tube to allow injection of large samples and to obtain complete conversion to ammonia. With this equipment they extended the useful range of the method to nitrogen levels below 1 ppm and to heavier stock. Good agreement (>90%) with the Kjeldahl method was reported for vacuum bottoms, lubricating oils and an olefin-amine copolymer with a molecular weight greater than 10 000. Low recoveries were obtained on the crude oils (35%) and asphaltenes.

With the aim of using the microcoulometric method for the determination of nitrogen in motor oil additives, Oita[(102)] modified the shape of the pyrolysis tube, and packed it with a hydrocracking catalyst (0·6% rhodium on silica–alumina) kept at 760°C and a hydrogenation catalyst (nickel-chips at 760°C and nickel on magnesia at 360°C). The samples were introduced in tin capsules. This system was used to determine nitrogen contents of motor oil additives, ranging from 44 ppm to 6%. The results were in good agreement with those obtained by the Kjeldahl method. With minor modifications, the method has also been used to determine nitrogen at the 1 ppm level.[(103)]

The introduction of heavy liquids and solids into the pyrolysis tube, preferably as solutions in a nitrogen-free solvent, is achieved conveniently by means of a boat inlet (Section III.F and Fig. 20). The inlet section of the combustion tube must be wide enough to accept the sample boat and must form a closed system with the boat attachment.

A determination is carried out as described in Appendix IV.8.

The data in Table 22 show the advantages of the boat inlet over the normal syringe injection for the analysis of heavy products and indicate good agreement with the Kjeldahl method. Prior to analysis the samples were diluted to about 50 ppm N with cyclohexane.

Table 22. Nitrogen in heavy materials determined by syringe injection and introduction with a sample boat.[98]

Sample	Kjeldahl (ppm N)	Microcoulometer (ppm N)			
		Syringe		Sample boat	
Light cycle oil	218	202	±10	190	±5
Gear oil	313	305	4	295	1
Motor oil 1	423	varies	±80	430	6
Motor oil 2	579	405	5	582	10
Gas engine oil	515	466	10	582	6
Transmission fluid	735	460	2	730	20
Industrial oil	769	634	4	758	18
Drawing oil	2500	2600	100	2300	100
Outboard motor oil	5600	varies	±2000	5400	100

The boat inlet was also successfully applied in the analysis of waste waters containing particulate matter.[98] With this technique it is possible to determine separately those nitrogen compounds which are volatile at room temperature, and total nitrogen. The analysis is carried out as follows. Inject the sample (5 µl) into the boat and push it near the pyrolysis tube inlet. Allow the water and other volatile components to evaporate and titrate. When the peak has returned to the baseline, push the boat into the inlet kept at 800°C and titrate the residual nitrogen.

Good agreement between the volatile nitrogen and the ammonia content of the wastewater samples as determined with the Nessler reagent indicated that the volatile nitrogen consisted predominantly of ammonia.[98]

Albert and co-workers[29] also adapted the microcoulometric system for the determination of total nitrogen and ammonia in municipal waste water. For this purpose, a specially designed sample inlet which includes a calcium oxide scrubber is placed between the pyrolysis tube and the titration cell, so that the catalyst can be by-passed for the determination of ammonia. The sample inlet and the scrubber are located in a heating block kept at 430°. The precision of the methods for total nitrogen and ammonia is within ±6% relative at levels of 5–50 ppm and the results compare well with those obtained by the Kjeldahl method.

Kajikawa *et al.*[104] determined the nitrogen content of crude oils and residual fuel oils by selecting the proper inlet temperature. The reproducibility of the results obtained on a heavy fuel oil is 2·8 ppm at the 27 ppm N level, and the accuracy is comparable to that of the Kjeldahl method.

G. Applications

The microcoulometric method for nitrogen has been tested on blends of organic nitrogen compounds in organic solvents[30, 64, 98] and in aqueous media and oxygenated solvents,[64, 98] as well as on inorganic salts.[29, 98] The method has also been applied to petroleum products ranging from light distillates to heavy products and residua[18, 19, 30, 98, 101–104] to nitrogen-containing pesticides and pharmaceuticals[99] and to municipal wastewater.[29, 98]

VIII. MAINTENANCE AND FAULT FINDING

The following recommendations should be taken into consideration when working with microcoulometric systems.

1. The whole system, starting from the gas supplies to the titration cell should frequently be checked for gas leaks, particularly when working with the reductive system, when careless handling of the hydrogen gas can cause explosions.

2. The microcoulometer module should not be placed on top of the furnace module, as its performance may be affected at elevated temperatures.

3. When the system is not in operation, an oxygen flow of 25 cm^3/min should be maintained through the pyrolysis tube and the cell in the case of the oxidative system, and a hydrogen or argon flow of 50 cm^3/min in the case of the reductive system.

4. With the reductive system, a vent should be attached to the cell outlet at all times to prevent hydrogen accumulation in the shielding cabinet which houses the cell and the instrument surroundings.

5. During catalyst conditioning and scrubber tube conditioning, the titration cell should be detached from the pyrolysis tube.

When attempting to locate a fault in the microcoulometric system, four major areas must be considered; the pyrolysis tube and furnace, the titration cell, the microcoulometer, and the recorder and/or integrator. To establish in which area the fault is, proceed as follows:

Fault Finding: List of Possible Faults, Diagnoses and Cures

Trouble	Diagnosis	Cure
Noisy baseline and/or low recoveries	1. Poor second-stage pressure regulators on gas supplies	Replace
	2. Gas leak somewhere between sample inlet and titration cell	Apply silicone grease to joints, replace septum if worn, check pyrolysis tube and titration cell for blockages
	3. Gas bubbles in the cell sidearms	Remove bubbles by flushing with electrolyte. In the case of reference sidearm, lift electrode, allow electrolyte to overflow and re-insert electrode (grease the joint).
	4. Electrostatic field around titration cell	Ground stirrer motor, place grounded Faraday cage around the cell
	5. Amplifier fault	Check first the two dry cells, clean contacts and replace cells if exhausted
	6. Recorder gain too high	Adjust gain
Noisy baseline and drift persist with constant-voltage source in place of the cell	1. Defective choppers in microcoulometer	Replace choppers
Poor precision	1. Worn or poor-quality syringe	Replace syringe
Cell not titrating	1. Electrodes not properly connected	Check connections between cell and microcoulometer for electric contact
	2. Faulty reference electrode	Inspect electrode, re-pack or replate if necessary
	3. Fault in microcoulometer	Refer to manufacturer's instructions
OXIDATIVE SYSTEM		
Noisy baseline and/or low recoveries	1. Contaminated or damaged cell cap electrodes	Remove cell cap, flame electrodes (S determination, Appendix I.1) or replate (Cl determination, Appendix II.5)
	2. Carbon deposits in pyrolysis tube inlet: temperature too high,	Remove tube from furnace, burn off deposits in a stream of air, check furnace

	outlet: sample too large, insufficient oxygen flow	
	4. Carbon deposits in cell capillary inlet: sample too large, insufficient oxygen flow	Drain electrolyte from cell, leave electrolyte in reference sidearm. Add few drops of conc. HF into capillary. Wash cell thoroughly with water, flush with electrolyte
	5. Quartz tube devitrification	Treat tube with 50:50 (by volume) hydrofluoric acid for 10 minutes. If trouble persists, replace tube
Baseline shifts with increasing attenuation (range ohms)	1. Impurities in gas supplies	Insert adsorber packed with activated 5A molecular sieve in line
Small explosion occurs when injecting sample	1. Leaky septum	Replace. Prevent pieces of old septum from getting into the pyrolysis tube
Peaks overshooting and baseline drifts downscale	1. Diffusion of electrolyte from reference sidearm into cell cavity due to leaky ground joint on reference electrode	Grease joint lightly and re-insert electrode.
Double or distorted peaks (Particularly in Cl determination)	1. Cell inlet heater not on	Check heater, replace if necessary
REDUCTIVE SYSTEM		
Low or variable recoveries, tailing peaks	1. Catalyst contaminated	Recondition catalyst (Appendices II.3, and IV.2)
	2. Pyrolysis tube coked	Recondition tube
	3. Restriction of gas flow	Check scrubber tube (N determination only), pyrolysis tube and titration cell for blockages
	4. Contaminated sensor electrode	Rinse with 10% HNO_3 (N determination only), replate if necessary (Appendices II.5 and IV.6 respectively)
Negative peaks	1. Scrubber exhausted or overloaded (N determination only)	Inject samples at slower rate. If trouble persists, re-pack or replace scrubber.

(a) Disconnect the pyrolysis tube from the titration cell and observe noise, background and recorder pen deflection. If the disturbances cease, the fault is caused by the tube or furnace, or the gases.
(b) Inject aqueous standards directly into the titration cell and evaluate the cell response. If the disturbances persist, the fault must be in the cell or in the electronics of the microcoulometer or the recorder.
(c) Substitute a constant voltage source (A-100 microcoulometer/recorder analyzer, Section III. G) for the titration cell. If this eliminates the disturbances, the fault is caused by the cell. If the disturbances persist, it can be assumed that the fault is in the microcoulometer or the recorder.

Consult the fault finding list for further identification of the cause of the trouble.

IX. DIRECT MICROCOULOMETRY

Microcoulometry can be applied directly for the determination of species which react with the titrants generated in the microcoulometric cells, if these species are already present as such in the sample to be analyzed, e.g. SO_2, hydrogen sulphide, thiols, HCl and ammonia. In this case, the decomposition step—combustion or hydrogenolysis—can be omitted and the sample introduced straight into the cell. If a sample is a gas or a sufficiently volatile liquid, it is injected into a carrier gas stream connected to the capillary inlet of the cell. If the sample is an aqueous solution, it is introduced by means of a syringe equipped with a long needle directly into the cell electrolyte. Although such methods are outside the general terms of reference of this book, it was considered useful to include a short account of these determinations for the sake of completeness.

In this field, continuous microcoulometric analyzers (Section III.C) have found their widest application, particularly in the analysis of gases. If the gas sample is passed through the cell at a constant flow rate, the concentration of the titrated component is directly proportional to the generating current required to restore the titrant ion concentration to the level, preset by the bias voltage. The sensitivity of the instrument is also a function of the flow rate; the higher the flow rate, the higher the sensitivity.

Bromine is a very convenient and therefore the most widely used titrant. The generating reaction at a platinum anode in acidic bromide media is predominantly

$$Br_3^- + 2e^- \rightarrow 3Br^- \qquad (37)$$

Hydrochloric or sulphuric acid added to the electrolyte furnishes the required high concentration of highly mobile cations for the transport of current through the cell. The current efficiency of bromine generation is practically 100%.[105]

In principle, continuous microcoulometers can also be applied to liquids; in this case, the solution to be titrated must be pumped at a constant rate past the sensor and then past the generator electrode.

Typical applications of direct microcoulometry follow.

A. Mustard gas in air

Shaffer, Briglio and Brockman[33] used a coulometer operating on the null-balance principle for the continuous monitoring of trace quantities of mustard gas (bis(2-chloroethyl)sulphide) in air. The cell, made of a 16 ounce glass bottle with a screw cap holding the electrodes, and the circuit are shown schematically in Fig. 30.

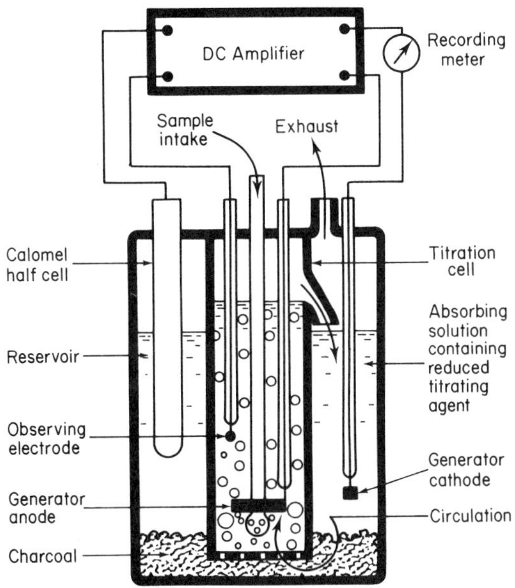

FIG. 30. Schematic diagram of titration cell and circuit for the continuous determination of mustard gas in air.[33]

The air is pumped at a constant rate through the electrolyte in the central compartment, consisting of 0·05 M potassium bromide in 3 M sulphuric acid. The mustard gas hydrolyses to thiodiglycol which is then oxidised to diglycol sulphoxide by the bromine generated at the platinum generator electrode:

$$(OHCH_2CH_2)_2S + Br_2 + H_2O \rightarrow (HOCH_2CH_2)_2SO + 2Br^- + 2H^+ \quad (38)$$

One mole of thiodiglycol (or mustard gas) is equivalent to two Faradays ($2 \times 96\,500$ coulombs). The potential of the platinum sensor (observing) electrode in the central compartment against a calomel reference electrode is presented to the amplifier, which returns an equivalent current to the generator electrodes to compensate for losses and to restore the initial very low bromine concentration in the electrolyte. The electrolyte circulates to the outer compartment where the excess of bromine is removed by passage through a bed of charcoal and re-enters the central compartment.

In addition to mustard gas, the analyzer can be applied to the determination of hydrogen sulphide, sulphur dioxide and acrolein down to a concentration of 0·1 ppm by volume.

B. Sulphur-containing gases in process streams and air

The Titrilog continuous analyzer (Section III. C), which has been developed from the mustard gas monitor of Shaffer *et al.*, has been used primarily to monitor sulphur dioxide, hydrogen sulphide, thiols, sulphides (thioethers), disulphides and thiophenes in process gas streams and air. The determination is based on the titration of these compounds with electrogenerated bromine which oxidizes sulphur dioxide to sulphate and hydrogen sulphide to sulphur. Thiols are oxidized to disulphides and thioethers to sulphoxides while carbonyl sulphide and carbon disulphide are not titrated. The instrument requires calibration for each sulphur compound. The electrolyte used in the cell is 0·1 M potassium bromide in 3·5 M sulphuric acid.

The threshold sensitivity of the instrument varies with the sample flow rate from 0·5 ppm by volume at 100 cm^3/min to 0·05 ppm at 1000 cm^3/min, the range (for hydrogen sulphide) is 0·05 to 120 ppm. The response time is about 30 seconds. Nader and Dolphin[(106)] improved the sensitivity of the instrument by a factor of more than 10 and McKee and Rollwitz[(107)] evaluated it in the field.

The response of the instrument is unaffected by saturated hydrocarbons, carbon monoxide, carbon dioxide, hydrogen, oxygen or ammonia. Olefins and diolefins react with bromine to a small extent which is, however, sufficient to cause a positive interference in the titration of sulphur compounds. Chlorine and nitrogen dioxide act like bromine and cause a negative interference.

Many attempts have been made to increase the versatility of such analyzers by extending their application to individual sulphur compounds with the use of selective absorption or adsorption. Tarman *et al.*[108] studied the efficiency and selectivity of solutions of various metal salts for the absorption of individual sulphur-containing gases, with the aim of applying coulometric analyzers for the analysis of odorants and other sulphur compounds in natural gas. The solutions were tested with the Titrilog and another coulometric analyzer at a gas flow rate of 500 cm^3/min, using samples of natural gas containing about 3 mg S/litre in the form of hydrogen sulphide, tert-butyl mercaptan (2-methyl-2-propanethiol), diethyl sulphide, thiophane (tetrahydrothiophene), or tert-butyl disulphide. The results are given in Table 23.

Table 23. Efficiency of absorbing solutions for the removal of individual sulphur compounds.[108]

Absorbent	Hydrogen sulphide	*t*-Butyl mercaptan	Diethyl sulphide	Thio-phane	*t*-Butyl disul-phide
	Percent sulphur removed				
Buffered cadmium sulphate	98	1	0	3	1
Alkaline cadmium sulphate	97	52–48	9	0	1
Alkaline cadmium sulphate + monoethanolamine	98	99	9	0	0
Potassium dichromate	98	28–15	0	2	0
Silver sulphate	100	99	99	99	1
Silver nitrate (16·8 g/litre)	100	100	100	100	0

These data show that all solutions tested were effective in removing H_2S. Alkaline cadmium sulphate with monoethanolamine was much more effective for combined H_2S and mercaptan removal than the alkaline cadmium sulphate solution as such. However, both solutions removed about 9% diethyl sulphide. Silver sulphate and silver nitrate solutions were equally effective in absorbing all sulphur compounds except the disulphide, but the sulphate was found to be less selective after 25–30 minutes of operation. The buffered cadmium sulphate solution showed good selectivity for H_2S in the presence of mercaptan, while the selectivity of dichromate was poor.

Tarman *et al.* also pointed out that the zero level current of the Titrilog increases with decreasing gas molecular weight. The slope of the line is close to that expected for critical flow as function of molecular weight.

The Titrilog is provided with a programming system which routes the sample through liquid absorbers, bypasses the absorbers and establishes a zero signal (baseline), using an automatic repetitive cycle. As an example of a typical absorption sequence, the sample is first directed through no absorber, then through an H_2S absorber and finally through an H_2S plus mercaptan absorber. The first recorded level represents total sulphur, the following level total sulphur less H_2S and the final level total sulphur less H_2S and mercaptans. The recorder pen then returns to the baseline and the system is ready to start the next cycle. The quantities of H_2S and mercaptans are obtained by simple subtraction. Each step of the cycle takes 15 minutes. A typical Titrilog chart is shown in Fig. 31.

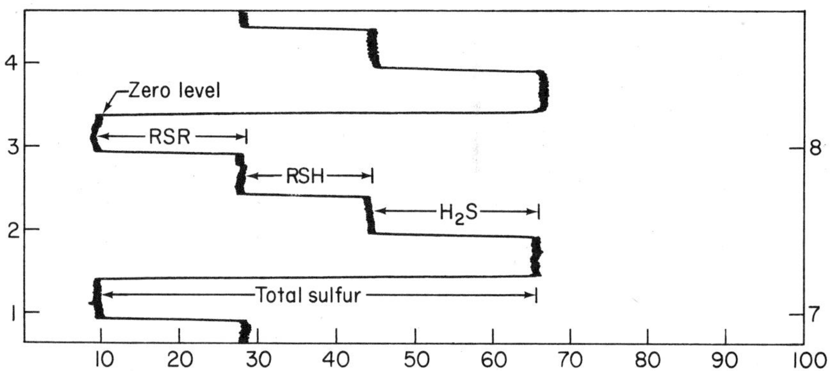

FIG. 31. Typical Titrilog chart, showing the total sulphur content of a sample and sulphur present as hydrogen sulphide, thiol (RSH) and sulphide (RSR), determined by selective absorption of the individual components.[71]

Jensen *et al.*[38] studied the absorption of hydrogen sulphide and methyl mercaptan by aqueous solutions of chlorine, sodium hydroxide, and chlorine plus sodium hydroxide. They found that aqueous chlorine solutions at a pH above 12 were effective absorbents for hydrogen sulphide but appeared to be impractical for methyl mercaptan since dimethyl disulphide, the major product formed in the oxidation, was stripped from the solution by the gas stream. Sodium hydroxide was an effective absorbent for both methyl mercaptan and hydrogen sulphide.

Smejkal and Bartonek[109] analyzed sulphur emissions from Kraft pulp mills, using selective absorption in solutions and coulometry. The analyzed gas is washed consecutively with the following solutions:

Absorbing solution	*Removes*
Potassium hydrogen phthalate	SO_2
1% Cadmium sulphate in 2% boric acid	$SO_2 + H_2S$
10% Sodium hydroxide	$SO_2 + H_2S + CH_3SH$
0·5% Silver nitrate	$SO_2 + H_2S + CH_3SH + (CH_3)_2S$ leaves only $(CH_3)_2S_2$ in sample gas

The method was applied successfully for the determination of sulphur dioxide, hydrogen sulphide and methyl mercaptan, but further separations especially that of dimethyl sulphide from dimethyl disulphide, were inadequate.

Pate and co-workers[110] found that membrane filters which had been impregnated with 5% potassium bicarbonate and then dried, quantitatively retained sulphur dioxide from air samples. Adams *et al.*[111] enlarged on this work and developed a method for the consecutive determination of sulphur dioxide, hydrogen sulphide, methyl mercaptan (MSH), dimethyl sulphide (DMS) and dimethyl disulphide (DMDS) in ambient air, by using a series of chemically impregnated membrane filters and microcoulometric determination of the sulphur compounds remaining in the air after passage through the filters. The analyses were carried out with a Dohrmann microcoulometer equipped with a bromine cell (Section III. C), using a constant air flow rate of 170 cm^3/min.

To prepare the filters, Type RA Millipore filters with a pore diameter 1·2 μm are soaked in the appropriate solution for 5–30 minutes, the excess liquid is drained off and the filters are dried on glass plates. The impregnated filters are then placed in filter holders and inserted in the air stream before the titration cell. The impregnants selected for the analysis and the retention of individual sulphur compounds achieved in percent are given in Table 24.

Table 24. Impregnants selected for the selective retention of sulphur compounds on membrane filters[111]

	pH	SO_2	H_2S	MSH	DMS	DMDS
1. $NaHCO_3$	7–11	100	10	4	5	3
2. Zinc chloride + boric acid	4·7	0	100	0	5	5
3. Ag membrane	—	0	100	100	0	0
4. Mercury (II) nitrate + tartaric acid	1·0	0	100	100	85	10
5. Silver nitrate + boric acid + tartaric acid	2·0	0	100	100	100	100

Best results are achieved with the filters placed in the order shown in Table 24. A sodium bicarbonate back-up filter is used with the mercury(II) nitrate and silver nitrate filters to eliminate interference with the microcoulometric cell, caused by the nitric acid liberated in the absorption reaction. Similarly, a sodium bicarbonate filter is placed before the zinc chloride-boric acid filter to prevent interaction with SO_2. The detection limits for the above sulphur compounds in air were found to be 25 parts per billion for sulphur dioxide, 10 ppb for hydrogen sulphide, 15 ppb for methyl mercaptan, 25 ppb for dimethyl sulphide and 5 ppb for dimethyl disulphide.

Bamesberger and Adams[112] reported on the application of the method to the determination of the above sulphur compounds in the atmosphere and measured the breakthrough time for each of the five filters. All filters met a minimum endurance requirement of at least 24 hours for sulphur gases in the range 50–100 parts per billion. Interference limits for common sulphur-free air pollutants were also studied and found to be negligible in most cases. The authors also compared the microcoulometric method with colorimetric and lead acetate tape methods for the determination of sulphur containing gases in the vicinity of a kraft pulp mill.[113]

Garber *et al.*[35] evaluated the Titrilog and Barton coulometric analyzers (Section III. C) for the monitoring of hydrogen sulphide evolved from sewage and waste water treatment plants and found the latter instrument to be the most accurate and versatile for overall use. The sensitivity of the Barton titrator for hydrogen sulphide is 0·05 parts per million by volume at the standard flow rate of 250 cm^3/min. Operation of the instrument is satisfactory from 200 to 500 cm^3/min.

Thiols and hydrogen sulphide in gases and volatile liquids can also be determined microcoulometrically by direct titration with electrogenerated silver ions. Both sulphur species form insoluble precipitates with silver ions, but two electrons are involved in the titration of H_2S while only one must be transferred for the reaction with RSH. Drushel[85] has used a special vaporization tube with two arms, one of which is packed with glass wool which has been immersed in a 0·1% aqueous solution of cadmium acetate and air dried. Both arms are equipped with septa for sample injection and nitrogen carrier gas inlets. The device is wrapped with a heating tape and kept a 240–250°C, and the joint outlet is attached to a T-300-S (or a T-400-S) silver cell (see Fig. 18). A carrier gas flow of 25–50 cm^3/min is maintained through both arms. A sample is injected with a syringe into the empty arm. The peak obtained represents the sum of H_2S and thiols. When a sample is injected into the packed arm, the packing will retain the hydrogen sulphide but allow the thiol to pass through. Thus, the peak obtained represents the thiol sulphur only. The sulphur response obtained from a vaporization through the arm packed with cadmium acetate is about 10% less than the response obtained in the empty arm.

The use of selective absorbers or adsorbers has now been largely superseded by gas chromatographic separation of the individual sulphur compounds, followed by coulometric or flame-photometric detection (see next Section).

Sulphur dioxide in combustion gases is readily determined by the injection of discrete samples into a carrier gas stream and direct microcoulometric titration in an iodine cell. Other constituents of the gas such as carbon monoxide, CO_2, SO_3 and HCl do not affect the titration. Chlorine and oxides of nitrogen interfere, but this interference is eliminated by the addition of sodium azide to the electrolyte unless these gases are present in excessive concentrations.

Microcoulometric analyzers are available, specially designed for the continuous monitoring of sulphur dioxide in ambient air at the parts per billion level. The Philips PW 9700 SO_2 Monitor is probably the most highly developed and most modern of these analyzers. A flow diagram is shown in Fig. 32. Automatic calibration and zero checking is achieved by means of a

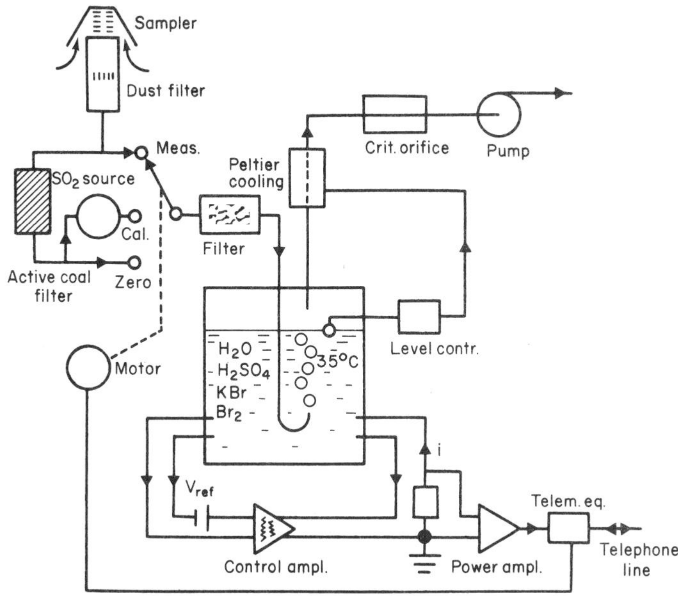

FIG. 32. Flow diagram of Philips PW 9700 SO_2 Monitor. (Courtesy of Philips Inc., Eindhoven, Netherlands).

remote-controlled three-way valve. With the valve in the "zero check" position, the air from the intake flows through a charcoal filter to remove SO_2. The signal from the coulometric cell then represents the zero level (baseline). With the valve switched to the "calibration" position the air passes through the same charcoal filter, then past an internal source of SO_2 and then to the measuring cell. This internal source supplies a constant flow of sulphur dioxide per unit time and thus provides a calibration. With the valve in the third, the "measurement" position, the charcoal filter is bypassed and the air proceeds directly to the coulometric cell, passing only through a silver gauze heated to 120°C, which eliminates any hydrogen sulphide, ozone or chlorine that would otherwise interfere with the SO_2 determination.

The titration cell is similar to that of the Titrilog analyzer (Fig. 16); it consists of an inner compartment where the reaction takes place, which accommodates the sensor electrode and the generator anode, and a larger vessel filled with electrolyte in which the reference electrode and the generator cathode are placed. The electrolyte in the inner compartment is connected to the outer vessel by a small aperture. The titration cell is kept at a temperature of 35°C at which bromine volatilization is not excessive. However, water evaporates from the electrolyte and as soon as the level falls below a certain limit, a Peltier cooling element fitted around the air outlet of the cell is automatically activated, cooling the air and causing the water vapour to condense and return the cell. The electrolyte is a solution of potassium bromide in dilute sulphuric acid. Air flow through the cell is kept constant at 150 cm^3/min by means of a suction pump and a thermostatically controlled critical orifice.

Stevens[(37)] who reviewed colorimetric, coulometric, conductometric and flame photometric techniques for the measurement of ambient concentrations of sulphur dioxide, observed the following response characteristics with the Philips monitor:

Minimum detectable level	0·01 ppm SO_2
Zero drift	less than 0·01 ppm per day
Precision (if instrument is calibrated daily)	better than 0·01 ppm
Response time to 90% of full scale (1·0 ppm)	2·5 minutes

Table 25 lists interferences by other air pollutants, expressed by the relation $I = Sx/S$, where S is the signal produced by 0·5 ppm SO_2 and Sx is the signal produced by a blend of 0·5 ppm SO_2 and 0·5 ppm of pollutant.

The feature described, in particular the automatic zeroing and calibration with a built-in source of sulphur dioxide, and maintenance of the electrolyte at a constant level make the analyzer one of the most trouble-free, drift-free monitors for sulphur dioxide available. It is designed to operate without maintenance for three months and is also equipped with a telemetering module for automatic transmission of readings to a central control station.

The use of permeation tubes is a simple way of preparing accurate low concentrations of gaseous air pollutants for the calibration of instruments. Liquefied gas is sealed in Teflon tubes with glass or metal bulbs. The gas permeates through the walls of the tube at a rate dictated by the geometry of the system and the properties of the compound. At constant temperature

Table 25. Interferences caused by other air pollutants in the measurement of sulphur dioxide with the Philips SO_2 monitor[37]

Pollutant	Interference
Nitrogen dioxide	−0·05
Nitric oxide	0·00
Ozone	0·00
Ethylene	0·02
Hydrogen sulphide	0·00
Methyl mercaptan	1·8
Benzene	0·00
Chloroform	0·00
Carbon disulphide	0·00
Propionaldehyde	0·01

the permeation rate is constant. Temperature is a critical variable that must be controlled within $\pm 0{\cdot}1°C$. Scaringelli *et al.*[84] used the Dohrmann microcoulometer to evaluate Teflon permeation tubes for sulphur dioxide.

O'Keeffe and Ortmann[114] described the preparation, calibration and use of permeation tubes for a variety of gases and vapours. Scaringelli *et al.*[84] and Cedergren *et al.*[115] used microcoulometry and the West–Gaeke colorimetric method to evaluate Teflon permeation tubes for sulphur dioxide. On the other hand, Rodes and co-workers[36] used gravimetrically calibrated permeation tubes to evaluate the performance characteristics of conductometric, colorimetric and coulometric (Barton) analyzers for the monitoring of sulphur dioxide in air.

Niklas[116] patented selective cartridges for the removal of hydrogen sulphide, chlorine and nitrogen dioxide from sample gas streams, which interfere in the coulometric determination of sulphur dioxide. Silica gel (1-mm particles) is soaked in a 4% aqueous solution of sulphamic acid and the wet grains are treated with a solution of 60 mg of sodium 4-(dimethylaminoazobenzene)-4-sulphonate in 100 ml distilled water. The air-dried material is packed into a glass tube. Another part of the tube or another tube is packed with silica gel impregnated with the sulphamic acid solution and then soaked in a copper sulphate solution. Sulphamic acid reacts with nitrogen dioxide, chlorine reacts with the azo dye and $CuSO_4$ removes hydrogen sulphide. The impurities are removed with better than 99% efficiency, without appreciable alteration of the SO_2 concentration in the gas sample.

Scaringelli and Rehme[117] determined atmospheric concentrations of sulphuric acid aerosol by microcoulometry. In this method the aerosols are collected by impaction or filtration, which separates the sulphuric acid from SO_2. The sulphuric acid is then isothermally decomposed to sulphur trioxide in a stream of nitrogen which is then converted to SO_2 over a bed of copper kept at 500°C and determined by titration with electro-generated iodine. The method can measure sulphuric acid in the presence of 10 to 100 times the amount of sulphur dioxide or other sulphates. The sensitivity of the method with a microcoulometric finish is 0·03 μg H_2SO_4, with a colorimetric finish 0·3 μg and with a flame photometric finish 0·003 μg per sample. Killer[133] discussed the microcoulometric determination of sulphur dioxide and other air pollutants.

C. Chlorides in aqueous media

Cedergren and Johansson[74] determined chloride in water down to 0·01 ppm by direct injection of the sample into the coulometric cell. The LKB 16300 Coloumetric Analyzer is used for the determination in conjunction with a cell with a rotating generator anode (see Section III. C). The electrolyte is 75% acetic acid; the error is 0·1–5% and depends on sample size.

Jacobsen and Tandberg[118] used a very simple battery-operated constant current source and biamperometric ("dead-stop") end-point detection for the same purpose. The electrolyte is 80% methanol containing 0·05 M nitric acid. The method which is simple and rapid gave satisfactory results in the range 0·1–100 ppm of chloride. In the concentration range 0·1–1·0 ppm Cl, the most accurate results were obtained when a known amount of chloride (1 μg per ml of water sample) was added before the titration.

Bromide, iodide, thiocyanate and cyanide interfere in the determination of chloride and must be absent from the solution. The interference from sulphite and nitrite can be eliminated by the addition of small amounts of hydrogen peroxide.[119]

Schait and Cuzner[120] used the Dohrmann microcoulometer to measure the chloride content of beer by direct injection of the sample into the coulometric cell.

Chloride titrations by means of silver ions must always be performed in acidic solutions in order to avoid the formation of silver oxide. Linnet[121] used an electrolyte consisting of 0·4 M sulphuric acid with 0·2% polyvinyl alcohol and a small amount of an antifoaming agent. The polyvinyl alcohol is added to keep precipitated silver chloride in suspension, thereby preventing a sluggish electrode response and a drop in current efficiency, caused by deposition of silver chloride on the generator and sensor electrodes.

When measuring the chloride content of aqueous media by straight injection into the cell electrolyte, always the same volume of standard and sample must be injected to cancel out the dilution effect which causes an error due to end-point displacement (Section VI. C). No gas flow through the cell is required for this determination; however, it is advantageous to maintain a nitrogen flow of 100–150 cm^3/min, as it enhances agitation of the electrolyte, thus producing sharper peaks, and prevents access of air and dust particles. The gas tubing should be of copper or polyethylene to eliminate even traces of contaminants.[68]

The method can be used to determine water-soluble chlorides in light hydrocarbon process streams. For this purpose, an adequate volume of the sample (up to 500 ml) is extracted with a few ml of chloride-free water, kept slightly acidic by the dropwise addition of dilute nitric acid. An adequate volume of the extract (usually less than 250 μl) is injected into the electrolyte of a T-300-S silver cell and the chloride content of the extract derived from a calibration with hydrochloric acid or sodium chloride standards. As little as 50 ppb of soluble chlorides can be determined in this way.

Ott and Gunther[122] described an automated system for the determination of chloride in aqueous samples, utilizing an Auto Analyzer and microcoulometric titration. Krijgsman *et al.*[68] used a microcoulometer of their own design and a T-300-S cell to determine nanoequivalents of halides. The electrolyte was 80% acetic acid containing 1% of sodium perchlorate. The relative standard deviation for the determination of more than 5 neq. chloride was better than 1%.

D. Acids and bases in aqueous media

The same authors also applied their microcoulometer for the determination of nanoequivalents of acids in aqueous solutions by titration with electrogenerated OH ions. The electrolyte is 0·04 M sodium sulphate in redistilled water, free of CO_2 The bias voltage is selected to keep the hydrogen ion concentration at pH 6·00 and nitrogen is bubbled through the electrolyte to eliminate interferences from the atmosphere. A relative standard deviation better than 1% was obtained in the determination of more than 20 neq of acid.

The determination of ammonia in aqueous samples by microcoulometric titration with hydrogen ions in a T-400-H cell has already been described (Section VII. G). For this purpose the sample is vaporized, passes through an acid gas scrubber packed with sodium hydroxide on alundum or with calcium oxide and is titrated in the cell. Vaporization with a boat inlet is preferred in the case of samples with particulate matter or high inorganic salt concentrations, such as sea water and brines. Hydrogen or an inert gas (argon, helium or nitrogen) can be used as the carrier gas.

E. Nitrogen oxides in gases and air

Concentrations of nitrogen dioxide in air up to 5000 ppm (by volume) can be determined by continuous coulometry, based on the reaction

$$NO_2 + 2H^+ + 2Br^- \rightarrow NO + H_2O + Br_2 \qquad (39)$$

which forms the basis of the Mast 724-11 analyzer.[123] The electrode system consists of a platinum cathode made up of many turns of wire wound around a pillar, and of a platinum anode, consisting of a single turn of wire. Air at 140 cm^3/min is pumped into the cell by a constant-volume piston pump, where it meets the bromide solution which is pumped at 2·5 ml/h by a 1/60 rpm motor-driven pump. The solution passes through a narrow annulus in intimate contact with the electrodes. Application of a potential across the electrodes causes a current to flow which is measured by the microcoulometer.

Scaringelli *et al.*[124] determined ppb concentrations of nitrogen dioxide, nitric oxide and ammonia in air. Ammonia and NO are converted to NO_2 over a platinum gauze at 370°C and NO_2 is then measured with a modified Mast coulometric analyzer.

Krichmar *et al.*[125] measure NO and NO_2 in nitrogen and air by oxidation of NO to NO_2 and detection of the total NO_2 by oxidation of iodide to iodine in an electrolyte containing 0·1 M potassium iodide. Bypassing the oxidation step gives the NO_2 corresponding to the NO content of the sample. The recovery of iodine is only 50–80% of the stoichiometric value; the coulometer must therefore be calibrated with synthetic blends of the gases to take into account all the factors influencing the conversion and titration. The relative standard deviations of the results obtained are better than 15% in the range 0·1–1000 ppm NO_x. The oxidation is carried out in a U-shaped oxidizer, made of 3 mm i.d. glass tubing, with a 15 mm i.d. bulb at the exit end. A small volume (0·10–0·15 ml) of a saturated solution of potassium permanganate in 10% sulphuric acid is placed in the oxidizer: pushed by the carrier gas stream (nitrogen) it travels as a slug up to the bulb and returns to the starting point, wetting the walls of the tube with a thin film of absorbent. One charge of the oxidizing solution is sufficient for 4–5 hours of continuous operation.

The same set-up can be used to determine NO in the presence of sulphur dioxide: bypassing the oxidizer, SO_2 is obtained from the amount of iodine formed in the coulometric cell (NO does not react with the electrolyte); with the oxidizer in line, NO is oxidized to NO_2 and determined as described above while SO_2 is absorbed by the oxidizing solution.

Similarly, NO is determined in the presence of ammonia by passage of the sample through the oxidizer which oxidizes NO to NO_2 and retains ammonia in the acidic absorption liquid. Passage of another sample through a spiral combustion tube heated to 500°C and through the oxidizer produces a mixture of nitrogen oxides which is then titrated as described above.

F. Kjeldahl nitrogen with coulometric finish

Boström and co-workers[126] and Cedergren and Johansson[127] described a rapid coulometric method for the Kjeldahl determination of nitrogen, based on the digestion of 10 mg samples in special digestion tubes, dilution of the digestion products, and microcoulometric titration of an aliquot with electrogenerated hypobromite ion. The method can handle 20–30 samples per hour. For substances containing nitrogen in the percent range the relative standard deviations for eight different substances were 0·1–1%.

The electrolyte used for the microcoulometric titration is 1·5 M in potassium bromide and 0·075 M in sodium tetraborate ($Na_2B_4O_7 \cdot 10H_2O$), adjusted to pH 8·60 ± 0·05 with 2 M sulphuric acid. At this pH the electrogenerated bromine disproportionates into hypobromite and bromide:

$$2Br^- \rightarrow Br_2 + 2e^- \tag{40}$$

$$Br_2 + 2OH^- \rightarrow BrO^- + Br^- + H_2O \tag{41}$$

At higher pH the disproportionation will go further to bromate, at a lower pH there is no disproportionation. The hypobromite formed reacts with ammonia at the anode according to the equation

$$3BrO^- + 2NH_3 \rightarrow N_2 + 3Br^- + 3H_2O \tag{42}$$

The determinations were carried out with the LKB 16300 Coulometric Analyzer and a cell with a rotating platinum generator anode. The indicating system consists of a platinum sensor electrode and an Ag/AgCl reference electrode. The generator cathode is a platinum wire inside a glass tube with a sintered disc to separate it from the sample compartment. The electrolyte volume is 30 ml. The current efficiency of hypobromite generation obtainable with the coulometric titration system is practically quantitative up to currents of the order of 100 mA, thus permitting rapid titrations.

G. Coulometric determination of trace water with the Karl Fischer reagent

Because of its specificity, the Karl Fischer reagent has been widely used for the determination of low concentrations of water in various organic and inorganic media. The determination is based on the following reaction

$$C_5H_5N \cdot SO_2 + [C_5H_5N \cdot I]^+ I^- + C_5H_5N + H_2O \rightarrow$$

$$C_5H_5N\langle\begin{matrix} SO_2 \\ | \\ O \end{matrix} + 2C_5H_5NH^+ I^- \tag{43}$$

which in the presence of methanol proceeds to

$$C_5H_5N\langle\begin{smallmatrix}SO_2\\|\\O\end{smallmatrix} + CH_3OH \rightarrow C_5H_5N\langle\begin{smallmatrix}H\\ \\SO_4\cdot CH_3\end{smallmatrix}$$

From these equations and the equation for the anodic oxidation of iodide

$$2I^- - 2e^- \rightarrow I_2 \tag{45}$$

it is evident that one mole of water consumes one mole of iodine or two redox equivalents (faradays).

Cedergren[(128)] proposed a method where the iodine is generated coulometrically and the end-point, corresponding to a slight excess of iodine, is detected potentiometrically with a non-polarized platinum electrode. Samples of 1–500 μl containing 0·05–200 μg of water have been analysed with a standard deviation of 0·015 μg in the range 0·05–20 μg of water. The determination is carried out with the LKB 16300 Coulometric Analyzer and a specially constructed cell, consisting of three chambers: one for the auxiliary generator electrode, one for the sample, housing the generator anode and the sensor electrode, and one for the reference electrode. Electrolytic contact between the chambers is made via liquid junctions packed with asbestos. All electrodes are made of platinum. The cell is fitted into a brass block and cooled to about 7·0°C.

A Karl Fischer reagent is chosen which consists of 1·0 M pyridine, 0·1 M iodine and 0·6 M sulphur dioxide, dissolved in methanol. All reagents are of reagent-grade purity. This reagent composition enhances the rate of the main reaction and suppresses side reactions.

Barendrecht and Doornekamp[(129)] applied null-balance coulometry to the continuous determination of water by the Karl Fischer method. A continuous flow of liquid (e.g. of methanol containing less than 1% water) is supplied to the coulometric cell at a flow rate of 20–200 ml/h. Iodine is generated in the anodic compartment of the cell at a rotating electrode. The electrolysis current is controlled by an amperometric sensing system and electronic feedback, thus generating the exact amount of iodine per unit time with a current efficiency of 100%.

H. Miscellaneous

Electrogenerated hypobromite can also be used for the coulometric determination of cyanide. Gibbs and Palma[(130)] utilized constant-current coulometry with biamperometric endpoint detection to determine 0·5–10 meq of cyanide in aqueous samples with an average error of 0·36%.

Braman *et al.*[(131)] developed a continuous microcoulometric titration system for the monitoring of boron hydrides (boranes) in air. In this system, boranes are scrubbed out of air with electrolyte made up of 0·03 M potassium iodide in 0·05 M sodium bicarbonate and titrated with electrogenerated iodine. In the coulometric system used, the amperometric indication system and the generation current are operated alternately in a cycle and the number of cycles recorded is a measure of the amount of boranes present in the sample. Without this operation in cycles, amperometric indication is difficult to apply in automatic coulometers because of the signal induced in the indicating electrodes by the generating current. The method is capable of determining diborane and decaborane in concentrations as low as 0·2 ppm by volume. Acetone and peroxides and any other substances that react with iodine interfere in the determination. Hydrogen gas, when bubbled directly through the titration cell, caused the titrator to generate iodine.

The various aspects of microcoulometry in non-aqueous media are of particular interest for the analysis of petroleum products. Typical applications are, besides the determination of moisture with the Karl Fischer reagent already discussed, the determination of neutralization values of lubricating oils and of bromine numbers, which express the degree of unsaturation of hydrocarbons. Miyake[(132)] investigated the conditions for the coulometric determination of these parameters, using a constant current source (20 or 40 mA), potentiometric end-point detection, and a cell made up of two semi-cylindrical glass vessels clamped together, with an ion exchange membrane separating the two chambers. One chamber, equipped with a stirrer, serves as the titration compartment, while the other houses the auxiliary generator electrode. The ion-exchange membrane prevents the migration of ions from one chamber to the other.

The composition of the electrolyte holds the key to the success of non-aqueous coulometry. It is convenient to consider different electrolytes in the two chambers: the electrolyte in the titration chamber must be non-aqueous, to act as a solvent for the sample and as a source for the titrant ion, however, an aqueous solution can be used in the other chamber. The electrical resistance of both electrolytes should be as low as possible. This can be achieved by the addition of halide, perchlorate or a neutral tertiary ammonium salt to the electrolyte.

Thus, hydrogen ions required for the determination of the base value of lubricating oils are generated in the titration chamber from an electrolyte consisting of 0·3 M sodium perchlorate in a mixture of methanol and chlorobenzene (50 + 50 by volume), while a 0·2 M aqueous solution of the same salt is used in the other chamber. A glass–calomel electrode pair serve as the indicator electrodes and the generator electrodes are made of flat platinum wire spirals. Oil samples (0·2–5 g) are dissolved in 25 ml of chlorobenzene and 1-ml aliquots are taken for the titration.

The same general considerations apply for the determination of bromine numbers where the titration is carried out with bromine, generated coulometrically from a 0·1–0·2 M solution of potassium bromide in a mixture of glacial acetic acid and methanol, to which a small amount of water is added to assist dissolution of the KBr. Less polar solvents may be added to dissolve the samples, and acetate or perchlorate to increase the electrical conductivity of the electrolyte. The electrolyte in the auxiliary electrode chamber can be a 0·1 M aqueous solution of a strong electrolyte.

X. THE MICROCOULOMETER AS GAS-CHROMATOGRAPHIC DETECTOR

Gas chromatography is one of the most powerful techniques for the separation of complex mixtures. The ever-increasing need for the determination of trace constituents in such mixtures has resulted in the development and use of selective detectors for gas chromatography, which eliminate unwanted peaks from the chromatogram and thus simplify the qualitative and quantitative analysis of complex mixtures. A thorough discussion of GC detectors is given by David.[(164)]

Microcoulometry with and sometimes without a preliminary degradation step offers itself for the selective detection of sulphur, chlorine, nitrogen and phosphorus-containing species in the effluent from a chromatographic column. Its major advantages, compared with the other detectors, are linearity of response and specificity, its major disadvantages are slow response and a relatively low sensitivity.

The response of the microcoulometer is linear over three orders of magnitude and is not affected by changes in carrier gas flow. A unique feature of the microcoulometer is that, with the combustion or hydrogenolysis step in line, quantitative evaluation of the chromatogram is possible without the need for a calibration; the amount of the substance analyzed can be calculated from Faraday's law, by taking into account the integral of the signal and the percentage of the element determined in the compound.

The specificity of the microcoulometric methods for sulphur, chlorine, nitrogen and phosphorus has been discussed in the previous chapters. As the microcoulometric detector is practically insensitive to hydrocarbons and their oxidation or reduction products, a low background is obtained. This fact permits the use of temperature programming and of more volatile or less stable stationary phases.

Some of the separation efficiency achieved on the chromatographic column is again lost in the microcoulometric detector, due to remixing in the pyrolysis tube and due to the time required for absorption in the cell electrolyte and titration. This leads to peak widening which is particularly noticeable on the first narrow peaks of a chromatogram. The width of later peaks which take one minute or more to elute is increased by about 30%, as compared with an instant-response detector such as the flame ionization detector (FID). Furthermore, additional tailing of the peaks on the chromatogram will be caused by the fact that the processes taking place in the titration cell (absorption of the analysed species in the electrolyte, titration and regeneration of the titrant) are diffusion controlled.[53] To obtain a satisfactory separation, it is therefore necessary to use longer columns and lower initial column temperatures than in the case of fast-response detectors. Because of the relatively large total volume of the microcoulometric detector (up to 200 cm^3), it can be used only with packed columns, not with capillary columns which give superior separations.

The sensitivity of the microcoulometric detector for sulphur is about 1 ng S. It is sufficient to determine individual sulphur compounds in gases down to 0·2 ppm by volume, corresponding to 0·3 mg S/Nm^3. The detection limit in liquids is about 0·1 ppm S in a 10-mg sample.[53] Compared with the flame photometric detector of Chaney and Brody[46] for sulphur (and phosphorus) compounds, the microcoulometer lacks in sensitivity and speed of response, but is superior in linearity of response and freedom from interferences.

The detection limit for chlorine compounds is about 1 ng Cl,[60] based on a gas sample volume of 5–10 ml. The other commonly used selective detector for the detection of halogen compounds is the electron capture detector. It can be very sensitive, but has only a narrow linear range, requires a separate calibration factor for each halogenated compound and is easily overloaded. Taking into account its inherent shortcomings, the microcoulometer is probably the most selective and versatile detector for halogenated hydrocarbons.

The detection limit of nitrogen compounds is also of the order of 1 ng N,[27] which corresponds to about 0·2 ppm nitrogen in liquid samples of the usual size. Of other detectors for nitrogen compounds, the Coulson conductivity detector[47] is equal in selectivity but superior in sensitivity to the microcoulometer. Due to the fact that both detectors utilize a catalytic hydrogenolysis step to convert the organically bound nitrogen to ammonia, they both suffer from an extended response time and the problems related to the hydrogenolysis. The thermionic detector is also very sensitive for nitrogen, but its performance depends on many factors.

A. Improved coulometric detectors

Attempts have been made to eliminate or to reduce the main shortcomings of the microcoulometric detector—slow response and insufficient sensitivity, e.g. by reducing the working volume of the detector. Thus, Krichmar and Stepanenko[162] designed a detector, consisting of a capillary tube with an i.d. of 1 mm. The electrolyte (a 0·01 M neutral solution of potassium iodide) is supplied to the capillary at the constant rate of 1 mm^3/s through a side arm, carrying two platinum generator electrodes. In the capillary it meets with the column effluent which forms uniform bubbles interspersed between slugs of electrolyte, as it travels downwards through the capillary, past the indicator electrodes made of 0·1 mm platinum wire and located in the lower part of the capillary. The lower end of the capillary is widened and carries a sidearm to separate the gas from the electrolyte, which is then discarded. The reactive components of the column effluent are absorbed in the electrolyte during passage through the detector and titrated with electrogenerated iodine. The threshold sensitivity with iodine as the titrant ion is 10^{-4} g-equivalents/s, so that down to 0·01 ppm by volume of sulphur compounds in gases can be determined with the detector. The same authors have also proposed an apparatus using permeation tubes for the preparation of synthetic gas blends containing 0·1–100 ppm by volume of trace components, and a low-inertia microreactor for oxidation and reduction of sulphur compounds to be attached to the exit of the chromatographic column.[163]

Sevcik[136] uses the flame ionization detector (FID) as a microcombustion unit and passes the combustion products through a heated glass capillary into a coulometric cell. The volume of the connecting tube is 0·6 ml, which under the experimental conditions chosen represents a time delay between FID signal and coulometer of 0·2 second. The coulometric cell consists of two arms in the shape of the letter D, filled with electrolyte. The combustion products are introduced near the bottom end of the inclined arm, travel upwards reacting with the electrolyte, and escape through a venting hole at the top of the cell. This causes circulation of the electrolyte in the cell and the reacted electrolyte reaches the sensor electrode located in the vertical arm which signals the generator anode, located near the bottom of the inclined arm, to regenerate the spent titrant ion. The time delay for one complete electrolyte circulation through the cell volume is less than one second. The auxiliary generator electrode and the calomel reference electrode are located in separate compartments.

When potentiometric endpoint detection and an electrolyte consisting of 0·01 M Br_2/Br^- and 0·1 M Cu^{2+}/Cu^+ was used for the titration of SO_2 and chlorine formed in the FID combustion, the detector gave a threshold sensitivity of 8×10^{-9} g/s for sulphur and of 1×10^{-8} g/s for chlorine in the analysis of a mixture of dimethylsulphoxide and chlorobenzene. The response of the detector is a linear function of the amount of sulphur and chlorine present in the sample. Water formed in the combustion dilutes the electrolyte so that the original concentration drops to one half after eight hours, thus requiring frequent replacement of the electrolyte.

The response time of Dohrmann cells can be reduced and the sensitivity increased by operation in the so-called Mode II.[48] In normal operation (Mode I) the gas stream from the pyrolysis tube enters the titration cell at a point between the sensor electrode and generator anode, so that the cell responds to concentration changes in the whole volume of the vigorously stirred electrolyte. In Mode II, the cell cap is turned 90° so that the gas stream impinges directly on the sensor electrode, thus creating an analyte concentration at the electrode surface which is much higher than the concentration in the bulk of the electrolyte. As a result of this arrangement, much larger and sharper peaks are obtained than in Mode I, with a corresponding increase in background noise, nevertheless resulting in a better signal-to-noise ratio by a factor of 10. The parameters affecting the cell time constant in Mode II are amplifier gain, the location of the sensor electrode (the surface area of the electrode swept by the gas stream) and the sample residence time on the electrode, which is a function of the gas flow rate and the stirring rate. The sharp peaks obtained in Mode II operation are in fact due to overtitration caused by the high analyte concentration localized on the sensor electrode. The cell will ultimately back-titrate the excess of titrant ions, but at the low gain and slow stirring rate used in Mode II this process is extended over a period of 10 to 15 minutes and manifests itself merely as a slightly lower baseline. Thus, at the trace level, peak height instead of peak area can be used as the detection criterion;[60] however, the amount of analyte cannot be

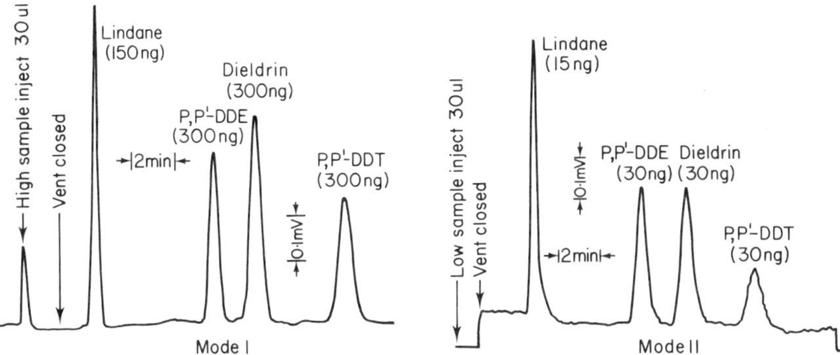

FIG. 33. Chromatograms of chlorinated hydrocarbons obtained with the T-300-S cell operating in Modes I and II[60]; Column: 1500 × 6.2 mm Pyrex glass, Packing: 1:1 mixture of 10% DC-200, 12,500 cs on 60/80 Chromosorb W-DMCS and 15% QF-1 on 80/90 Anakrom ABS; Temperatures: injection 240°C, column 235°, transfer 240°, furnace, center 900°, furnace, outlet 900°. Flow rates: carrier gas (argon) 100 cm^3/min, oxygen 65 cm^3/min.

calculated from the area of a peak obtained by Mode II operation by means of Faraday's law. The difference between the microcoulometer responses in Modes I and II is illustrated by the chromatograms in Fig. 33, which were obtained by the gas-chromatographic separation of a synthetic blend of the chlorinated hydrocarbons lindane, *p,p′*-DDE, dieldrin and *p,p′*-DDT, followed by combustion and titration of the HCl formed with silver ions in a T-300-S cell, using Mode I and Mode II operation.[60] The conditions for the chromatographic separation and combustion were the same; the microcoulometer settings were as follows:

	Mode I	Mode II
Gain	3600	600
Range (ohms)	44	200

Stirrer speeds as low as 40 rpm have been recommended for Mode II operation.[48]

B. Apparatus, materials and operating conditions

A gas chromatograph, coupled to the microcoulometric system for elemental analysis described in the previous chapters is required for the analysis of sulphur, chlorine, nitrogen or phosphorus compounds in complex mixtures. Microcoulometric detector systems are available,* consisting of the pyrolysis unit, the coulometric cell and the microcoulometer, which are specially adapted for use with a gas chromatograph.[137–140] They have been designed to provide quick response and high sensitivity, by reducing the working volume of the detector to a minimum, by avoiding losses in tubes and packings through the use of glass and Teflon as the major building materials, etc. The flow scheme of the GTS-20 detector for chlorine is shown in Fig. 34.

The chromatographic column is connected by Teflon tubing to a glass T-piece, one arm of which leads to the vent line and the other (again through Teflon tubing) to the pyrolysis tube. A toggle valve is installed at the end of the vent line to ease venting of the solvent. The pyrolysis tube is made up of three quartz tubes that fit into each other as shown on the diagram. Provision is also made for the column effluent to be split between the microcoulometric and another detector, if desired. A simple automatic venting valve for Coulson conductivity and microcoulometric detectors has also been proposed.[141]

The same gases, packings and electrolytes are used in the microcoulometric detector as in the microcoulometric system for total elemental analysis described in the previous chapters. The chromatographic column materials must be selected for a particular application and will be discussed in detail

* Dohrmann, Envirotech, Mountain View, Cal.

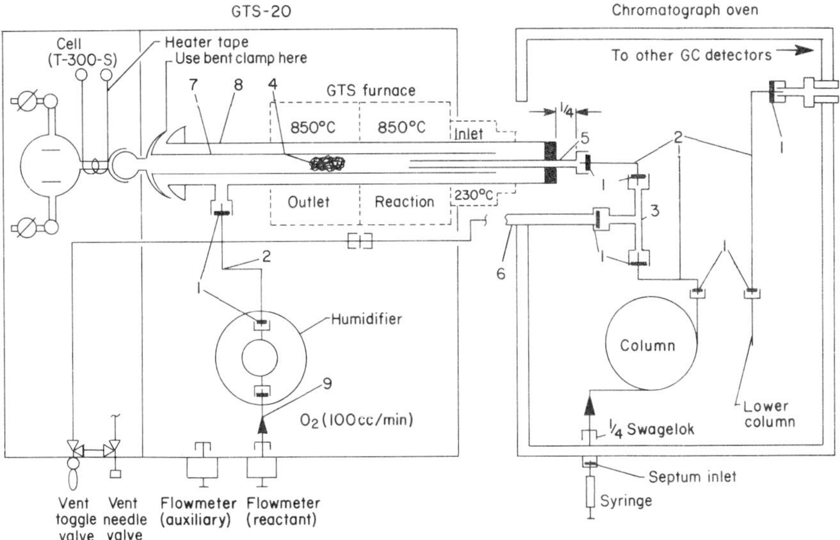

FIG. 34. Flow scheme of GTS-20 microcoulometric detector system (Dohrmann): (1) Septum connectors, (2) 1/16″ Teflon tubing, (3) Pyrex T-piece, (4) Quartz wool plug, 15 mm, (5) Column effluent tube, (6) 1/8″ stainless steel vent line, (7) Reaction tube insert, quartz, (8) Reaction tube, quartz, (9) 1/8″ plastic tubing.[138]

when discussing the applications of GC-microcoulometry. The column should not absorb any of the components irreversibly.

Standards should be chosen so that they have chromatographic characteristics similar to the expected sample components. Variations in recovery from the chromatographic column and the pyrolysis unit may require a separate standard for each sample component measured to obtain quantitative results.

The operating conditions of the detector differ to a certain extent from those used for total elemental analysis. These differences are dictated mainly by the fact that gas flow rates must satisfy not only the requirements of the microcoulometer but also those of the gas-chromatographic separation, that a smaller volume of the pyrolysis unit results in shorter residence times, etc. The recommended operating conditions for the four main versions of the microcoulometric detector are collected in Table 26.

The concentration range over which the microcoulometric detector can be applied is approximately 0·1 to 500 ppm of the element determined.

The sample is injected on the column, with the vent opened to allow the solvent and other components of the sample which elute before the components of interest to escape through the vent. This avoids overloading of the

Table 26. Operating conditions recommended for the microcoulometric detector.[137–140]

Element detected	Sulphur	Sulphur	Chlorine	Nitrogen
Pyrolysis	Oxidative	Reductive	Oxidative	Reductive
Gas flow rates (cm^3/min)				
Column carrier gas	Helium	Hydrogen	Helium, argon	Hydrogen
Oxygen (humidified)	100	—	100	—
Hydrogen (humidified)	—	100	—	100
Auxiliary gas (10% NH_3 in N_2)	—	40	—	—
Furnace temperatures (°C)				
Transfer zone	230	230	230	230
Reaction zone	850	1000[a]	850	700[b]
Outlet zone	850[c]	1000	850	300[d]
Cell model	T-300-P	T-400-S	T-300-S	T-400-H
Titrant ion	I_3^-	Ag^+	Ag^+	H^+
Microcoulometer				
Bias voltage (mV)	160	110	250	100
Gain (approx)	200	400	2400	400
System recovery[e] (%)	85 ± 10	100 ± 5	≥80	100 ± 5

[a] Packed with 3 g of platinum-on-alundum catalyst.
[b] Packed with 7·6 cm of nickel coil catalyst.
[c] Contains 15 mm quartz wool plug.
[d] Packed with 10/20 mesh K_2CO_3.
[e] Calculated from instrumental data using Faraday's law.

pyrolysis unit with unwanted material. After most of the solvent has passed through the column, the vent is closed and subsequent sample peaks swept into the pyrolysis unit where they are either combusted in oxygen or hydrogenolysed over a catalyst with a secondary stream of hydrogen. The products reach the titration cell and are titrated coulometrically as described previously. Quantitation is achieved through calibration by means of adequate standards and/or by calculation from electrochemical data.

C. Applications of gas chromatography with a microcoulometric detector

The applications of this combination of analytical techniques can be grouped into three major groups: (1) the determination of sulphur and nitrogen compounds in petroleum products, (2) the determination of pesticide residues in various media, such as soils and biological materials, and (3) the determination of various pollutants in the atmosphere.

1. *Direct determination of thiols in petroleum products*

Several authors have determined thiols in natural gas and low boiling petroleum fractions by direct titration of the individual thiols eluted from the chromatographic column with electrogenerated silver ions, thus eliminating the pyrolysis step. Except in the case of Liberti and Cartoni, the T-300-S cell (Fig. 18) was used for the titrations.

As mentioned in the introduction, Liberti and Cartoni[1] were the first to combine gas chromatographic separation with microcoulometric detection to determine C_3–C_5 alkanethiols in synthetic blends and gasoline fractions. Satisfactory separations were obtained on a glass column packed with silicone DC 550 on Celite. The cell electrolyte used for the determination was 75% ethanol containing 0·2 M $NaNO_3$ and 0·05 M perchloric acid.

Fredericks and Harlow[10] used an aluminium column packed with tricresyl phosphate on Chromosorb-W to determine individual thiols in natural gas containing large amounts of hydrogen sulphide. To eliminate the H_2S, the column is vented through an Ascarite absorber until the hydrocarbon constituents and the H_2S are eluted at room temperature. Individual thiols are then determined by temperature programming of the column and microcoulometric titration with silver ions. The precision of the method is ±2% at 50 ppm and the sensitivity is about 1 ppm of individual thiol. Sec-butyl and isobutyl mercaptans are only partly resolved.

Brand and Keyworth[142] separated thiols in the gasoline boiling range on a stainless steel column packed with di-*n*-decyl phthalate on Chromosorb-P. Recoveries better than 90% are obtained, as established on synthetic blends.

Drushel[85] obtained good resolution between H_2S and the lower thiols on a glass column packed with Porapak Q porous polymer beads, thus eliminating a liquid substrate. However, the column tends to release substances which react with silver ions and produce a high background in the region where C_4 and C_5 thiols are eluted. H_2S and thiols are partially retained by the column, resulting in low recoveries which in the case of thiols are of the order of 60–70%.

The operating conditions used in the above chromatographic separations of thiols are collected in Table 27.

1. *Determination of sulphur compounds in petroleum fractions*

A glass column (6 ft × $\frac{1}{4}$ in) packed with Porapak Q 80/100 mesh and maintained at 130°C can be used to determine hydrogen sulphide, COS and methanethiol in hydrocarbon gases. The products are eluted from the column with helium at 30 cm^3/min, combusted with oxygen and the SO_2 formed in the combustion titrated with electrogenerated iodine. The method is sensitive to sulphur compounds below 1 ppm.

Ripperger[14] used a column (10 m × 3 mm i.d.) packed with 7·5% polypropyleneglycol on Chromosorb-G to determine thiols in natural gas, hydrogen sulphide, thiols and thiophene in naphtha, and hydrogen sulphide

Table 27. Operating conditions for the chromatographic separation of thiols

Column	Dimensions	Packing	Carrier gas	Flow rate (cm^3/min)	Temperature (C°)	Ref.
Glass	1200 × 4 mm i.d.	40% silicone DC 550 containing 10% stearic acid on Celite.	N_2	30–80	78°	(1)
Aluminium	25 ft × $\frac{1}{4}$ in o.d.	30% TCP on Chromosorb W (acid washed)	He	120	ambient to 130° at 2·5°/min	(10)
Stainless	6 ft × $\frac{1}{4}$ in o.d.	30% didecyl phthalate on 60/80 Chromosorb P (acid washed)	He	50	60 to 160° at 5°/min	(142)
Glass	6 ft × $\frac{1}{4}$ in i.d.	Porapak Q 80/100 mesh	N_2	30	5 min at 75°, 75 to 225° at 5°/min	(85)

and tetrahydrothiophene (THT, used as stenching agent) in town gas. The products are eluted from the column, temperature-programmed from 50 to 150°C at 2°/min, with nitrogen at 25 cm^3/min, combusted at 750°C in a quartz tube filled with quartz chips and titrated as SO_2 with electrogenerated iodine.

Krichmar and Stepanenko[(134)] separated low-reactivity sulphur species in coke still gas, such as COS, carbon disulphide and thiophene on a column (1 m × 4 mm i.d.) packed with 20% silicone oil on Chromosorb-W and kept at 50°C, with air as the carrier gas. The sulphur compounds in the effluent are burned in a microreactor kept at 1000°C and titrated with electrogenerated iodine in the microcoulometric detector cell (Section X.A).

Klaas[(4)] determined sulphur compounds in naphthas at the parts per million level by chromatographic separation, followed by combustion of the sulphur compounds in the chromatographic effluent to SO_2 and coulometric titration of the SO_2 with bromine. The set-up consists of two chromatographic columns, a fraction collector (cold trap), the combustion unit and a Titrilog continuous analyzer (Section III.C), modified to increase its sensitivity by a factor of 20. Two gas chromatographic columns, one polar, the other nonpolar, are used to achieve both a type and molecular weight separation of the sulphur compounds. Fractions are collected from the effluent of the polar column in a stainless steel trap, cooled with liquid nitrogen, and subsequently introduced into the nonpolar column by rapid heating of the trap to +100°C.

The polar column is made of stainless steel tubing 1 ft long and $\frac{1}{2}$ in diameter, followed by a tube 80 ft long and $\frac{1}{4}$ in diameter, packed with 30% tetrahydroxyethylethylenediamine (THEED) on 40/60 mesh Chromosorb. The column temperature is 120°C and helium is used as the carrier gas. The second (nonpolar) column is also made of stainless steel (18 ft × $\frac{1}{4}$ in) and packed with 30% silicone oil DC-200 on Chromosorb. This column is operated at 160°C. Separate pressure and temperature controls on the two columns are essential for continuous operation, since the retention times on the second column must be short enough to permit development of each fraction before the next fraction from the first column is introduced.

The method separates alkyl sulphides, cyclic sulphides and thiophenes. Identification of the various sulphur compounds is based on retention times on the polar and nonpolar columns. Thiols are strongly adsorbed on the THEED column and are not observed. The minimum detectable amount is about 1 ppm of sulphur as a single compound in a 100-μl sample.

Martin and Grant[(11)] studied the distribution of sulphur compounds in light and middle petroleum distillates and crude oils by gas chromatographic separation, combustion of the eluted sulphur compounds and titration with electrogenerated iodine, using a Dohrmann microcoulometer. The separations are made with a stainless steel column, 20 ft by $\frac{3}{16}$ in i.d., packed with 15% SE-30 methyl silicone on 30/60 mesh Chromosorb W (acid and caustic washed). The column temperature is programmed at 4°/min from 60 to as high as 400°C. The carrier gas is nitrogen or helium at 100 cm^3/min.

The nonpolar silicone column was chosen because it elutes sulphur compounds nearly in order of boiling point, regardless of compound type. The column begins to bleed at about 300°C, but this does not cause a significant response from the coulometric cell. The relatively high coating of silicone rubber is used to avoid adsorption of thiols on the Chromosorb support. Individual sulphur compounds down to about 5 ppm S can be determined only in the lower boiling fractions. Typical chromatograms, showing the distribution of sulphur compounds by boiling point in a gasoline (total sulphur content 0·090% S) and a coke-still naphtha (0·72% S) are presented in Fig. 35. In the gasoline most of the compounds are thiophenes, a few of which are identified by their retention times (thiophene, 2- and 3-methylthiophene and benzo/b/thiophene). The others are identified only by carbon number. The breakdown of the sulphur compounds into thiols, sulphides, disulphides, thiophenes and benzothiophenes, determined by other methods, confirmed the abundance of thiophenes in the sample. In the coke-still naphtha which is formed from residuum by thermal cracking, the distribution of sulphur compounds is more complex. Thiophenes still predominate, but

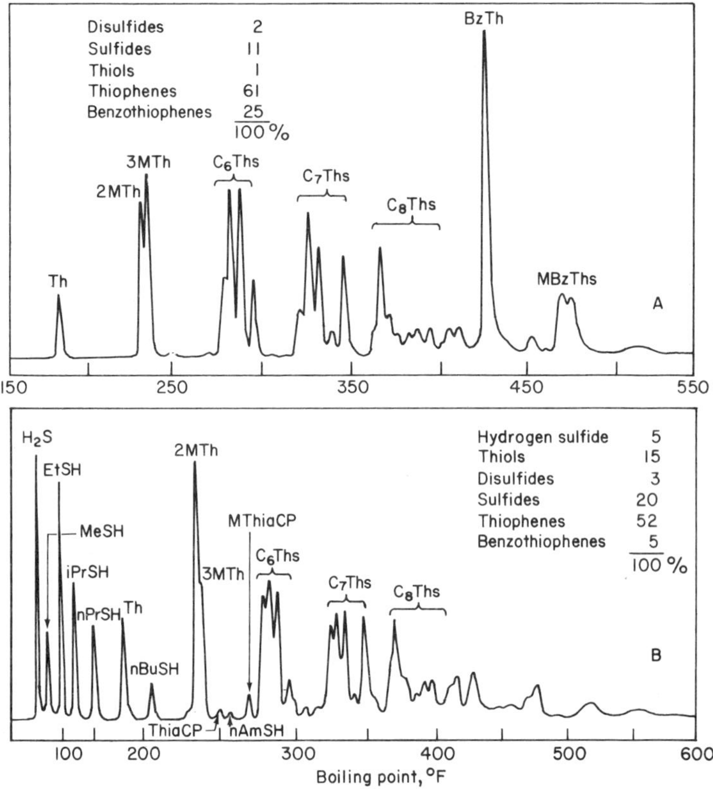

FIG. 35. Sulphur compounds by boiling point in (A) gasoline, (B) coke still naphtha.[11]

significant amounts of hydrogen sulphide, thiols and sulphides are also present.[11]

Schulz and Munir[53] identified more than 60 sulphur compounds in natural gas, crude oils and naphtha fractions, above all by gas chromatography with a microcoulometric detector, using combustion of the sulphur compounds to SO_2 and coulometric titration with iodine. The separations are carried out mainly on a column of Apiezon-L (10 m × 3 mm) with nitrogen as the carrier gas at 25 cm^3/min. After 15 minutes at 50°C the column temperature is raised to 250°C at 0·5°/min. The column effluent is mixed with another 25 cm^3/min of nitrogen and introduced into a quartz tube packed with quartz chips where it is combusted with 100 cm^3/min of oxygen at 800°C. The combustion products reach the cell where they are titrated to give a signal proportional to the sulphur content of the fraction in the usual way.

The authors identified 30 sulphur compounds in natural gas in the range C_1 to C_5 and found that they consisted mostly of thiols. In a hydro-desulphurized naphtha the lower thiols were removed by the treatment, and the remaining sulphur compounds were mainly aliphatic and cyclic sulphides. A comparison of the composition of thiols and thioalkanes in crude oil led to the conclusion that a genetic relationship exists between those classes of sulphur compounds.

Martin and Grant[(12)] developed a method for the determination of thiophenic compound classes in petroleum products. Nonthiophenic sulphur compounds (mostly aliphatic sulphides) are decomposed in a nitrogen atmosphere over alumina at 500°C to form hydrogen sulphide and aromatic thiols. These are collected in an aqueous sodium hydroxide solution and titrated potentiometrically with silver nitrate solution, to give total nonthiophenic sulphur. Thiophenic compounds, extensively dealkylated on the alumina, are trapped in a hydrocarbon solvent, separated by gas chromatography according to number of rings in the molecule and detected by microcoulometric titration with iodine after combustion to SO_2. The column used for the chromatographic separation (8 ft $\times \frac{1}{8}$ in i.d.) is packed with 3% diethyleneglycol sebacate polyester (LAC-737) on Chromosorb-W. The carrier gas is nitrogen at 140 cm^3/min, and the column temperature is programmed from 50 to 280°C at 2°/min.

Since the cracking on alumina dealkylates the thiophenes sufficiently to prevent overlap of the different classes, thiophene types with one, two, three, four and five or more rings can be distinguished by the method. The recovery of the thiophenes from alumina is practically complete. The fate of compounds which are higher boiling than naphthobenzothiophenes is uncertain; even if removed from the alumina they are too high boiling to be determined by the gas chromatographic procedure.

Drushel[(16)] made the above method simpler and more rapid by including a pyrolysis unit (Fig. 36) in line before the gas chromatographic column, thus achieving dealkylation, separation and determination of thiophenes in a single operation. The sample is pyrolyzed in the all-glass pyrolysis tube at 750°C in a stream of nitrogen (50 cm^3/min). The products are swept directly onto a chromatographic column also made of glass, through the combustion unit and finally into the coulometric titration cell where the SO_2 formed is titrated with electrogenerated iodine. Typical sulphur chromatograms of a gas oil (400–425°C fraction) obtained with a polar column (10% Carbowax 20 M on Chromosorb W) and a nonpolar column (10% ethylene–propylene copolymer on Chromosorb W) are shown in Fig. 37. The nonpolar column shows good resolution between individual substituted thiophenes, but the polar column gives better resolution by molecular type.

A pyrolysis temperature of 750°C was chosen because preliminary experi-

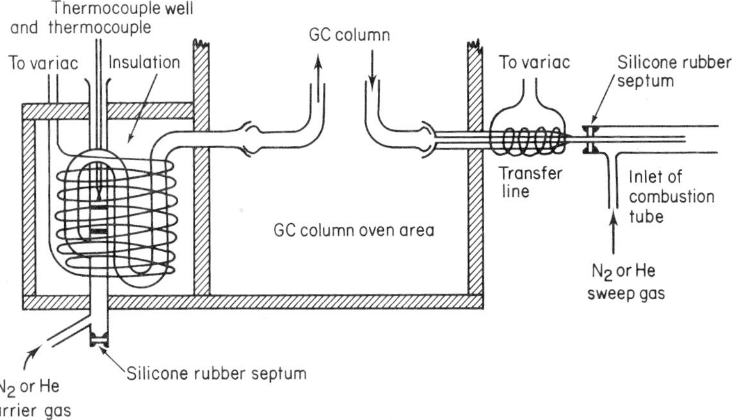

FIG. 36. In-line pyrolysis-gas chromatography-combustion system.[16]

ments showed that thiophene, benzothiophene and dibenzothiophene survived pyrolysis at this temperature with little change, while *n*-hexyl sulphide (representing aliphatic sulphides) decomposed to give H_2S with 98–100% conversion. At higher temperatures an increasing amount of sulphur is retained in the carbon deposits formed in the pyrolysis unit; furthermore,

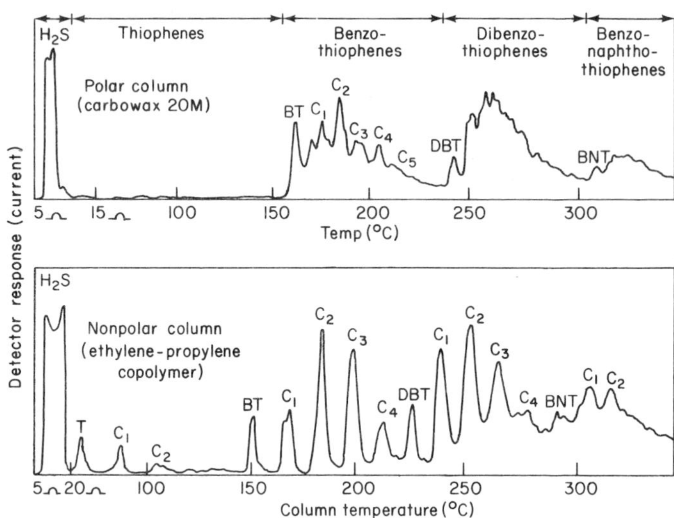

FIG. 37. Sulphur chromatograms of a gas oil obtained by the combined pyrolysis-GC-microcoulometric method.[40]

thiophenic compounds begin to decompose, producing measurable amounts of H_2S. In the case of gas oils, the amount of sulphur remaining in the pyrolysis unit at 790°C is about 30% of the total sulphur.

3. *Determination of nitrogen compounds in petroleum fractions*

Because of their harmful effect on processing and product quality, characterisation of nitrogen compounds by types and boiling points is particularly important in the petroleum industry. To meet this need, Martin[27] studied the conditions for the gas chromatographic separation of these compounds, subjecting the column effluent to catalytic hydrogenolysis to convert the nitrogen compounds to ammonia, and titrating the ammonia with electrogenerated hydrogen ions in a T-400-H cell. A column packing which gave efficient separations, did not react with or adsorb nitrogen compounds and which could be taken to relatively high temperatures without causing a detector response or catalyst deactivation was found in 9% polyethylene of 12 000 molecular weight on Chromosorb W, precoated with 3% potassium carbonate. The column (20 ft × $\frac{3}{16}$ in i.d. stainless steel) can be used up to 330°C which is sufficient to elute samples boiling in the gas oil range.

Because of the slow response of the microcoulometric detection system, sharp and incompletely separated chromatographic peaks cannot be followed perfectly; time delays and mixing in the catalyst tube and the titration cell cause broadening which can be sometimes avoided by operating the column at slow rates of carrier gas flow and temperature programming. This problem disappears, however, as retention times increase and peaks become wider.

The method has been applied to a catalytic cycle oil containing 260 ppm and a shale naphtha containing 1920 ppm of nitrogen. Both analyses were made with the temperature programmed at 1·4°/min, the cycle oil from 120 to 330°C and the naphtha from 50 to 190°C. The carrier gas (hydrogen) flow rate was 200 cm^3/min for the cycle oil and 120 cm^3/min for the naphtha. Pyridines, indoles, carbazoles and naphthobenzopyrroles dominate in the cycle oil, with some overlap between the compound types. The nitrogen compounds in the shale naphtha are almost exclusively substituted pyridines.

Albert[31] combined the method with chemical separation methods and mass spectroscopy to determine quantitatively the nitrogen compound distribution in light catalytic cycle oil and light virgin gas oil. Pyridines and quinolines are extracted from benzene solution with hydrochloric acid, indoles with 60% perchloric acid, and carbazoles and phenazines with 72% perchloric acid.

Drushel[32] used the pyrolysis–GC combination described above for the separation of sulphur compounds to study the distribution of nitrogen compounds in various petroleum fractions before and after hydrogenation. The column (3 ft × $\frac{1}{4}$ in i.d.) is made of glass, packed with 12% ethylene–

propylene copolymer on 80/100 mesh Chromosorb W, specially pretreated with potassium carbonate to neutralize acid sites. The packing withstands operation to 350°C without serious degradation and does not poison the granular nickel hydrogenolysis catalyst, used to convert the nitrogen compounds in the column effluent to ammonia in a stream of hydrogen at 800–850° C.

Prior to application of the method, basic nitrogen compounds are separated from the nonbasic nitrogen compounds by extraction with 20% hydrochloric acid to avoid mutual interference of the two compound classes in the chromatographic separation. A T-400-H cell is used to titrate the ammonia formed in the hydrogenolysis.

The method has been applied to the determination of nitrogen compound types in the raffinates and basic nitrogen extracts of light catalytic cycle oils, hydrogen-treated shale oils and others. Pyrroles and intermediates, indoles, carbazoles, benzcarbazoles and dibenzcarbazoles are distinguished in the nonbasic nitrogen fraction (raffinate) and pyridines and intermediates, quinolines, acridines and benzacridines in the basic nitrogen fraction (extract). The recovery of the nitrogen compounds is about 80–90%.

4. *Determination of pesticide residues*

The most widely used analytical technique for the determination of pesticide residues in soils, foodstuffs and other biological materials is gas chromatography, coupled with element-specific detectors: microcoulometric detectors for chlorine and sulphur compounds, flame photometric and thermionic detectors for sulphur and phosphorus compounds, electron affinity detectors for chlorine and nitrogen compounds, and others. The gas chromatographic approach is preferred as it does not require the more elaborate clean-up of spectroscopic methods and because it is more sensitive than most colorimetric methods. High pressure liquid chromatography is also employed particularly for the analysis of thermally unstable pesticide residues that pose problems in gas chromatography.

Microcoulometry as the detecting technique, following gas chromatographic separation, has found many applications in the analysis of pesticide residues. Its major advantages are linearity and stoichiometry of response which allows quantitation of peaks without cumbersome calibrations. Its performance in this field has been compared with that of the electrical conductivity detector by Cranmer and Carroll(143) and with the electrical conductivity and electron capture detectors by Westlake.(144)

Following the early work of Coulson and his co-workers(2,3) on chlorine pesticides which brought the microcoulometer into life, Burchfield and co-workers(5,6,17) developed systems for the separation and identification of sulphur, chlorine and phosphorus-containing pesticides in extracts from a variety of materials. They combined gas chromatography on polar (silicone QF-1 mixed with silicone SE-30) and nonpolar (DC-200 silicone oil) columns with the selective detection of sulphur and chlorine by oxidation and titration of the HCl and SO_2 formed in a T-300-S silver cell and a T-300-P iodine cell, respectively. Alternatively, phosphorus and sulphur are detected by reduction in an empty quartz tube at 950°C in a stream of hydrogen and titration of the PH_3 and H_2S formed in a T-300-S silver cell. If the effluent from the reduction tube is passed directly into the titration cell, PH_3 and H_2S (and HCl if present) are measured simultaneously with a relative sensitivity of 2:2:1. If a short tube packed with alumina is inserted between the reduction tube and the titration cell, H_2S and HCl are removed quantitatively while PH_3 passes unchanged and is titrated with silver ions according to the equation

$$2Ag^+ + PH_3 \rightarrow Ag_2PH + 2H^+ \tag{46}$$

Phosphorus and sulphur can be measured in the presence of one another by inserting a short silica gel column between the reduction tube and the titration cell which retains HCl irreversibly. The recovery of pesticides by the oxidative and reductive methods is about the same, in the order of 70%. Burchfield and co-workers also applied the method to study the metabolism of the tranquilizing drug chlorpromazine.[17]

Nelson[13] determined organic thiophosphate pesticide residues on fruits and vegetables by gas chromatography with microcoulometric sulphur detection. The Mills–Onley–Gaither procedure for the extraction and cleanup of chlorinated pesticide residues[145] was adapted and the extract injected into the chromatographic column. Two columns are used for the separations: a 4 ft column packed with 5% of DC-200 silicone on acid-washed Chromosorb P, and a 6- ft column packed with 15% of the same stationary phase. The temperature of the first column is 190°C, and of the second 240°C; nitrogen is used as the carrier gas, at a flow rate which elutes parathion in 6–8 minutes. The column effluents are burned in an empty quartz tube at 800°C and the SO_2 formed is titrated in a T-300-P iodine cell. The thiophosphates studied were Systox, Delnav, Diazinon, Di-Syston, methyl parathion, parathion, malathion, ethion, Trithion, EPN and Guthion. Only the parent compounds were investigated, no attempt being made to study the metabolites and decomposition products. Satisfactory recoveries are obtained, ranging from 68% for Delnav to 107% for ethion, except for Systox which gives only a recovery of 33%.

Aavik and co-workers[57] studied the factors affecting the chromatographic separation and microcoulometric determination of pesticides containing both sulphur and chlorine. The separations are carried out on a column 2 m × 4 mm i.d. packed with 20% polyethyleneglycol 4000 on Chromosorb

P with hydrogen at 40 cm^3/min. Placing an empty quartz tube heated to 900°C in front of the column, the HCl and H_2S formed in the hydrogenolysis of the chlorine and sulphur compounds are either determined simultaneously, on a 3·5 m glass column packed with Polysorb I, or separately, by selective adsorption of HCl on silica gel or porous glass, or of H_2S on mercurous iodide. The species passing through the system is titrated coulometrically in a silver cell with 60–70% acetic acid as the electrolyte. Recoveries of sulphur and chlorine-containing pesticides based on the two elements are about the same and range from 86 to 108%. The sensitivity of the method (for chlorine) is about 1 ng.

The system can also be used to determine organic phosphorus compounds.[58] In this case the reduction is carried out in an empty quartz tube at 950°C. The phosphine formed in the reaction is titrated with silver ions as Ag_2PH, using 35% acetic acid as the electrolyte. The recovery based on phosphorus is about 70%; the sensitivity is 5 ng P.

Pease and Kirkland[146] determined residues of the insecticide methomyl (S-methyl-N-[(methylcarbamoyl)oxy]thioacetimidate) in animal and plant tissues and soil. The method is based on extraction of methomyl from the substrate with ethyl acetate and subsequent alkaline hydrolysis to the volatile more stable oxime, methyl-N-hydroxythioacetimidate:

$$CH_3{-}S{-}C(CH_3){=}N{-}OCONHCH_3 \xrightarrow{OH^-} CH_3{-}S{-}C(CH_3){=}N{-}OH \tag{47}$$

The oxime is extracted with an organic solvent and the extract injected on a chromatographic column (4 ft × $\frac{3}{16}$ in i.d.) made of glass, packed with 10% FFAP on 80/100 mesh Chromosorb W, acid-washed and treated with dimethyldichlorosilane before use. The carrier gas is helium at 100 cm^3/min, and the column temperature is programmed from 100 to 200°C at 7·5°/min. The column effluent is passed through a combustion unit and the SO_2 formed titrated in a T-300-P cell with electrogenerated iodine. The sensitivity of the method is about 0·02 ppm based on a 25 g sample, with an average recovery of 93% for the tissues investigated. Somewhat lower recoveries are obtained with soils.

Takahashi *et al.*[60] studied the separation and detection of the organochlorine insecticides lindane, aldrin, dieldrin, *p,p′*-DDE, and *p,p′*-DDT, using a chromatographic column (1500 × 6·2 mm) made of glass, packed with a 1:1 mixture of 10% DC-200 on 60/80 Chromosorb W-DMCS and 15% QF-1 on 80/90 Anakrom ABS. The carrier gas is argon at 100 cm^3/min, the column temperature 235°C. The column effluent is passed through a combustion tube kept at 900°C and the HCl formed is titrated in a T-300-S cell with electrogenerated silver ions. With the cell operated in Mode II (see Section X.B), less than one nanogram of chlorine can be detected. The superior selectivity of the microcoulometric detector, as compared with the electron capture detector, is demonstrated.

Iwata and co-workers[147] used the same chromatographic column and analytical set-up to determine residues of the above organochlorine insecticides in raw and canned beef and cheese fat. The clean-up procedure of Wood[148] based on the selective solubility of these pesticides in dimethyl sulphoxide, was employed to prepare the extracts for GC-microcoulometry. Recoveries not lower than 60% but generally greater than 80% were obtained.

Organochlorine pesticides in human blood were determined by Griffith and Blanke[149] using gas chromatographic separation on OV-210 or SE-30/QF-1 columns followed by combustion and microcoulometric detection of the HCl by titration with silver ions. A modified sulphuric acid method is used to extract pesticides and industrial chemicals from human whole blood. The method allows the specific determination of as little as one part per billion of pesticides such as lindane and aldrin.

Munro[150] determined 2,4-dichlorophenoxyacetic acid and its 2,4,5-trichlorohomologue in tomato plants and other commercial crops by GC-microcoulometry. Intensive clean-up, including extraction, conversion of the free acids to the methyl esters and distillation, allows determination of 0·5 μg of the pesticide in 100 g of tomato plant with a recovery of 72–96% for the dichloro derivative and of 94–100% for the trichloro derivative.

Pease[151] determined residues of the weed killer bromacil (5-bromo-3-sec-butyl-6-methyluracil) in soil and in plant and animal tissues by extraction with a sodium hydroxide solution and partition into an organic solvent. Gas chromatographic separation of the extract on a stainless steel column (2 ft $\times \frac{3}{16}$ in i.d.), packed with SE-30 silicone plus 0·2% Epon Resin 1001 on 60/80 mesh Diatoport, (carrier gas helium at 75 cm^3/min, column temperature 300°C) is followed by combustion of the column effluent with oxygen and titration of the HBr with electrogenerated silver ions. Taking into account incomplete conversion of bromine compounds to a titratable species (only 40–50% of the organic bromine reaches the cell as HBr, see p. 176), by frequent calibration, average recoveries of 85% are obtained, at a sensitivity of 0·04 ppm, based on a 25 gram sample.

Aavik and co-workers[99] studied the catalytic hydrogenolysis of organic nitrogen compounds, the chromatographic separation of their degradation products and the microcoulometric titration of ammonia formed as the major hydrogenolysis product. Using a nickel catalyst for the conversion, recoveries of 98–100% were obtained for nitrogen-containing pesticides such as parathion and metaphos and drugs such as penicillin. The sensitivity of the method is about 1 ng N, with a selectivity coefficient of 10^5.

McCarthy[152] also used GC-microcoulometry with a hydrogenolysis step to determine metabolites of the insecticide carbofuran (2,3-dihydro-2,2-dimethyl-7-benzofuranyl-N-methylcarbamate) in plant and animal tissues by quantitation of the ammonia formed in the hydrogenolysis. The organic solvent extract obtained by the clean-up procedure is separated on a column packed with 20% SE-30 silicone on silanized Gas Chrom Z support. A T-400-H cell is used to titrate the ammonia with coulometrically generated hydrogen ions. The sensitivity of the method for corn grain is 0·05 ppm for a 70-gram sample, with recoveries of 85% for carbofuran and 79% for 3-hydroxycarbofuran, its major metabolite.

5. *Determination of pollutants in various atmospheres*

Gas chromatography combined with selective detectors is widely used to meet the ever increasing need for the determination of trace contaminants in various atmospheres. Here again, the microcoulometric detector has found many applications, particularly in cases where selectivity and stoichiometry of response outweigh its relatively low sensitivity and slow response. Thus, GC-microcoulometry has been used successfully to determine sulphur compounds in the gaseous effluents from kraft pulp mills which, although present in very low concentrations, are recorded by the human nose as an unpleasant odour. These compounds include primarily sulphur dioxide, hydrogen sulphide and alkyl mercaptans, sulphides and disulphides.

Bromine is the obvious titrant of choice for the coulometric titration of these sulphur compounds. When used directly to monitor the chromatographic column effluents, it is necessary to calibrate the cell response for each compound type because of the different electron requirements of the above sulphur compounds. Alternatively, a combustion or hydrogenation step can be included between column and cell to convert the effluent to SO_2 or H_2S, respectively. This eliminates the need for calibration and also eliminates possible interference from olefinic compounds which might be present in complex gas mixtures.

Jensen *et al.*[(38)] used a 20 ft $\times \frac{3}{16}$ in column of 20% Triton X-305 on Chromosorb-P for the separation and a Titrilog continuous analyzer (p. 186) to titrate the sulphur compounds. Adams and Koppe[(153)] increased the sensitivity of the method by replacing the Titrilog cell with a T-300-P cell, modified by Adams *et al.*[(73)] (see Section III. C) to operate with bromine instead of iodine as the titrant. A similar chromatographic column is used for the separations (8 ft $\times \frac{3}{16}$ in stainless steel, packed with 10% Triton X-305 on 60/80 mesh Chromosorb-G, DMCS treated, column temperature programmed 30° to 70°C at 10°C/min, carrier gas helium at 50 cm^3/min). The system is calibrated with blends of the pure sulphur compounds in nitrogen, prepared in plastic bags.

Smith and Tauss[(154)] applied GC-microcoulometry for monitoring gaseous pulp mill emissions to cover a concentration range from several parts per billion to several percent (as H_2S). The separations are obtained on an open-tubular column of 10% Triton X at 25° and 60°C and on a polyphenylene ether/Halopore column at 60°C.

Robertus and Schaer[(155)] developed a portable continuous GC-micro-coulometric analyzer for the quantitative analysis of sulphur compounds in stack gases. The instrument which operates in 10-minute cycles, again uses a bromine cell to titrate the sulphur compounds in the chromatographic column effluent and can detect 1 ppm of SO_2, 0·2 ppm of H_2S and 0·1 ppm of CH_3SH.

Brink *et al.*[(156)] designed a dual-flame ionization-microcoulometric titration system and used it to detect sulphur-containing and sulphur-free compounds that were produced in the pyrolysis of kraft-black liquor and eluted from a gas chromatographic column. The separations are carried out on a 10 ft $\times \frac{1}{8}$ in glass column packed with 20% DC-710 on 60/80 mesh Chromosorb W, acid-washed and silanized prior to use. Following sample injection, the column is first held isothermally at 50°C, then programmed from 50° to 200°C at 2·5°C/min and held at this temperature to completion of the run. The carrier gas is nitrogen at 20 cm^3/min. The column effluent is split 4:1 between FID and microcoulometer. Hydrogen at 90 cm^3/min is added to the microcoulometer stream, and the sulphur compounds are reduced at 1100°C to H_2S, which is then titrated with silver ions in a T-300-S cell. The FID response is recorded simultaneously; the delay due to the dead volume of the microcoulometer is approximately 24 seconds. Compounds identified include methyl mercaptan, dimethyl sulphide, dimethyl disulphide, the lower aromatics, phenol, anisol, cresols and xylenols.

Krichmar and Stepanenko[(134)] determined sulphides and disulphides in air by separation on a silicone oil column at 60°C, followed by combustion at 1000°C and titration of the SO_2 formed with electrogenerated iodine.

Schait and Cuzner[(120)] determined dimethyl and diethyl sulphides in beer at the parts per billion level. The beer head space samples are chromatographed on a 5 ft $\times \frac{1}{8}$ in stainless steel column packed with 25% tricresyl phosphate on 60/70 mesh acid-washed Chromosorb W. The column temperature is kept at 80°C for 5 min, programmed from 80 to 110°C at 4°/min and kept at this temperature until completion. The carrier gas is helium at 20 cm^3/min. The column effluents are reduced in a stream of hydrogen at 1100°C and the H_2S formed titrated with silver ions in a T-300-S cell. Dimethyl disulphide is used as the internal standard. Attempts to concentrate and chromatograph mercaptans from beer head space gave negative results.

Saalfeld *et al.*[(157)] discussed the problems involved in the determination of trace contaminants in enclosed atmospheres, such as those of spacecrafts and submarines. The complexity of such atmospheres is illustrated by the fact that, while 50–100 contaminants were identified in the early spacecraft atmospheres, the number of contaminant species is believed to be higher by a factor of 100 in nuclear submarine atmospheres.

Time-integrated adsorption of trace contaminants on charcoal is used to concentrate these components for identification and a variety of techniques is employed to analyze the desorbates, such as gas chromatography, micro-coulometry, mass spectrometry, infra-red spectroscopy and combinations of GC with the other techniques. The GC–MS and GC-microcoulometry combinations are the most useful. The latter is used primarily to determine chlorinated hydrocarbons of which dichloroacetylene (DCA) is a particularly toxic representative. A whole air sample is passed through a Porapak column maintained at 50°C to separate the contaminants from the large volume of air with which they are diluted. The column is then temperature programmed, the effluent is pyrolyzed at 1100°C, and the decomposition products are titrated with silver ions in a T-300-S cell. The sensitivity of this method, which has also been used to determine bromine and iodine compounds, ranges from 10 ppb to 10 ppm.

Williams[(158)] described an improved GC-microcoulometric method for the determination of DCA in complex atmospheres which takes advantage of the fact that certain column materials can stabilize the DCA during the separation; after the separation it is easily pyrolysed and the products titrated coulometrically. Such a chromatographic column packing is polyethylene glycol 400 (15% on Chromosorb P 80/100 mesh).

Williams and Umstead[(159)] took advantage of the fact that halogenated hydrocarbons have quite long retention times on porous polymer beads at room temperature, to collect chlorinated and brominated hydrocarbons in air on a column packed with Porapak Q and to elute them subsequently as symmetrical peaks by temperature programming as described above. Relatively large air samples (100 to 500 cm^3) can be used, and water does not interfere in the determination. Mixtures of various compounds containing Cl, Br and/or F (including Freons) are determined quantitatively down to concentrations of 20 ± 10 ppb. The base line of the microcoulometer is not affected by the temperature programming, although a slight degradation of the polymer beads occurs at high temperatures. Killer[(160)] discussed applications of the microcoulometric detector in pollution analysis.

6. *Miscellaneous*

High molecular weight chlorinated paraffins with 20 to 24 carbon atoms in the molecule and containing 40–70% chlorine are added to many industrial products as plasticizers, extenders and fire retardants. It is possible that these compounds leak into the environment in a manner similar to polychlorinated biphenyls.

To determine chlorinated paraffins in environmental samples, Zitko[161] applied chromatography on alumina and silica columns to prepare extracts, the chlorine content of which was then determined by conversion to HCl and microcoulometric titration with silver ions. Hexane and 10% diethyl ether in hexane are used as the eluents. The recovery of chlorine from the high-molecular-weight compounds is only 40–60% but can be made practically quantitative by the addition of 20% of di-(2-ethylhexyl)phthalate (DEHP) or Nujol to the hexane when preparing the solutions for microcoulometric chlorine determination.

XI. REFERENCES

1. A. Liberti and G. P. Cartoni, *Chim. Ind.* (Milan), **39** (1957), 821.
2. D. M. Coulson and L. A. Cavanagh, *Anal. Chem.* **32** (1960), 1245.
3. D. M. Coulson, L. A. Cavanagh, J. E. DeVries, and B. Walter, *J. Agr. Food Chem.* **8** (1960), 399.
4. P. J. Klaas, *Anal. Chem.* **33** (1961), 1851.
5. H. P. Burchfield, J. W. Rhoades, and R. J. Wheeler, *J. Agr. Food Chem.* **13** (1965), 511.
6. H. P. Burchfield and R. J. Wheeler, *J. Ass. Offic. Anal. Chem.* **49** (1966), 651.
7. J. A. Burke and W. Holswade, *J. Ass. Offic. Anal. Chem.* **47** (1964), 845.
8. J. A. Burke and L. Y. Johnson, *J. Ass. Offic. Anal. Chem.* **43** (1962), 8.
9. H. V. Drushel and A. L. Somers, *Anal. Chem.* **39** (1967), 1819.
10. E. M. Fredericks and G. A. Harlow, *Anal. Chem.* **36** (1964), 263.
11. R. L. Martin and J. A. Grant, *Anal. Chem.* **37** (1965), 644.
12. R. L. Martin and J. A. Grant, *Anal. Chem.* **37** (1965), 649.
13. R. C. Nelson, *J. Ass. Offic. Agr. Chem.* **47** (1964), 289.
14. W. Ripperger, *Gas u. Wasserfach,* **109** (1968), 786.
15. F. J. Wilby, Amer. Gas Assoc., Operating Section Proceedings, (1965), No. 65-P-137.
16. H. V. Drushel, *Anal. Chem.* **41** (1969), 569; Preprints *Div. Petr. Chem. ACS,* **13** (1968), No. 3, 121.
17. H. P. Burchfield, D. E. Johnson, J. W. Rhoades, and R. J. Wheeler, *J. Gas Chromatog.* **3** (1965), 28.
18. H. V. Drushel, Preprints, *Div. Petr. Chem. ACS,* **14** (1969), No. 3, B-232.
19. H. V. Drushel, *Anal. Letters,* **3**(7) (1970), 353.
20. F. C. A. Killer and K. E. Underhill, *Analyst,* **95** (1970), 505.
21. R. Moore and J. A. McNulty, Proceedings, 14th Annual Analysis Instrumentation Symposium, Philadelphia, May 1968.
22. L. D. Wallace and D. W. Kohlenberger, R. J. Joyce, R. T. Moore, M. E. Riddle, and J. A. McNulty, *Anal. Chem.* **42** (1970), 387.
23. F. C. A. Killer, *Erdöl und Kohle,* **23** (1970), 655.
24. H. A. Braier, J. Eppolito, and W. C. Zemla, *Preprints,* Div. Petr. Chem. ACS, **18** (1973), 72.
25. R. A. Hofstader, *Microchem. J.* **11** (1966), 87.
26. F. A. Gunther and J. H. Barkley, Bull. Experim. Contamination and Toxicology, 1966, **1**, No. 2, 39.

27. R. L. Martin, *Anal. Chem.* **38** (1966), 1209.
28. R. Moore and J. A. McNulty, "Determination of total nitrogen in water by microcoulometric titration". Paper presented at Pittsburgh Conference on Anal. Chem. and Appl. Spectroscopy, Cleveland, Ohio, March 1968.
29. D. K. Albert, R. L. Stoffer, I. J. Oita, and R. H. Wise, *Anal. Chem.* **41** (1969), 1500.
30. P. Gouverneur and F. van de Craats, *Analyst*, **93** (1968), 782.
31. D. K. Albert, *Anal. Chem.* **39** (1967), 1113.
32. H. V. Drushel, *Preprints,* Div. Petr. Chem. ACS, **14** (1969), No. 3,B-223.
33. P. A. Shaffer, A. Briglio and J. A. Brockman, *Anal. Chem.* **20** (1948), 1008.
34. R. R. Austin, Instrument Society of America, 1965 *Preprint* No. 2.1-1-65.
35. W. F. Garber, *J. Water Pollut. Contr. Fed.* **42**(5) (1970), R209.
36. C. E. Rodes, H. F. Palmer, L. A. Elfers, and C. H. Norris, *J. Air Pollut. Control Ass.* **19** (1969), 575.
37. R. K. Stevens, "Review of analytical methods for the measurement of sulphur compounds in the atmosphere". Presented at 11th Conference on Methods in Air Pollution and Industrial Hygiene Studies, California State Dept. of Public Health, March 1970.
38. G. A. Jensen, D. F. Adams, and H. Stern, *J. Air. Pollut. Control Ass.* **16** (1966), 248.
39. L. B. Ryland and M. W. Tamele. *In* J. H. Karchmer (Ed.), "The Analytical Chemistry of Sulphur and its Compounds", Part I, Wiley-Interscience, New York, 1970.
40. V. H. Drushel, *ibid.*, Part II, Wiley-Interscience, New York, 1972.
41. H. E. Aavik and V. Milli, "The application of gas chromatography with a microcoulometric detector in sanitary and hygienic studies". *In* Proceedings of the conference dedicated to 75 years of the Hygiene Dept. of the Tartu State University and 30 years of the Tartu Hygiene Epidemic Station, Tartu, SSSR, 1970.
42. J. A. Stamm, "Recent advances in applications of the microcoulometric titrating system", Lectures on Gas Chromatography 1966, Plenum Press, New York, 1967.
43. D. A. Leathard and B. C. Shurlock, "Identification Techniques in Gas Chromatography," Wiley-Interscience, London, 1970.
44. M. Krejci and M. Dressler, *Chromatog. Rev.* **13** (1970), 1.
45. D. F. S. Natusch and T. M. Thorpe, *Anal. Chem.* **45** (1973), 1185A.
46. S. S. Brody and J. E. Chaney, *J. Gas Chromatog.* **4** (1966), 42.
47. D. M. Coulson, *J. Gas Chromatog.* **4** (1966), 285.
48. J. A. McNulty and L. W. Hoppe, "Sub-micro Elemental Analysis by Microcoulometry", Dohrmann Instruments Co., Mountain View, Cal., 1969.
49. G. de Groot, P. A. Greve, and R. A. A. Maes, *Anal. Chim. Acta*, **79** (1975), 279.
50. A. Cedergren, *Talanta,* **20**, (1973), 621.
51. A. Cedergren, *ibid.* **22** (1975), 967.
52. J. P. Dixon, *Analyst*, **97** (1972), 612.
53. H. Schulz and M. Munir, *Erdöl und Kohle*, **25** (1972), 14.
54. W. W. Marsh, *Anal. Letters*, **3**(7) (1970), 341.
55. D. M. Coulson, *Amer. Lab.* 1969, (May), 22.
56. W. Walisch and O. Jaenicke, *Mikrochim. Acta (Wien)*, **6** (1967), 1147.
57. H. E. Aavik, A. V. Kabun, R. Kallasorg, and I. A. Revel'skii, "Study and development of a method of microcoulometric detection of halogen- and sulphur-

containing compounds", Tr. Vses. Isssled. Soveshch. Ostatkov Pestits. Profil Zagryazneniya Imi Prod. Pitan., Kormov Vnesh. Sredy, d, 1971, p. 16.

58. H. E. Aavik, R. A. Kallasorg, and I. A. Revel'skii, "Investigation of the conditions required for the microcoulometric determination of phosphor-organic compounds", *ibid.*, p. 23.
59. W. Lädrach, F. van de Craats, and P. Gouverneur, *Anal. Chim. Acta*, **50** (1970), 219.
60. Y. Takahashi, R. T. Moore, and M. E. Riddle, "A new microcoulometric detection system for gas chromatography: chlorinated hydrocarbon application". Presented at 84th Ann. Meeting, Ass. Offic. Anal. Chemists, Washington, DC, October 1970.
61. H. V. Drushel, "Determination of nitrogen in high-boiling fractions by microcoulometry", Presented at Southwest Reg. ACS Meeting, Baton Rouge, La., December 1972.
62. J. A. McNulty, *Amer. Lab.* (April 1969), 59.
63. Application Note MC-502: Quartz Exit Tubes, Dohrmann Envirotech, Mountain View, Cal.
64. D. R. Rhodes and J. R. Hopkins, *Anal. Chem.* **43** (1971), 630.
65. C-300 Microcoulometer, Instruction Manual, Dohrmann Envirotech, Mountain View, Cal.
66. LKB 16300 Coulometric Analyzer, LKB-Producter AB, Bromma, Sweden.
67. H. E. Aavik, V. Joonson, I. A. Revel'skii, A. V. Kabun, R. Kallasorg, P. Keres, V. Kindel, and A. Herem, "Microcoulometric detector for selective determination of halogen-, sulphur-, phosphorus-, and nitrogen-containing compounds", Tr. Vses. Issled. Soveshch. Ostatkov Pestits. Profil. Zagryazneniya Imi Prod. Pitan., Kormov Vnesh. Sredy, 1970, p. 11.
68. W. Krijgsman, W. P. van Bennekom, and B. Griepink, *Mikrochim. Acta,* 1972, 42.
69. W. Krijgsman, G. de Groot, W. P. van Bennekom, and B. Griepink, *ibid.,* 1972, 364.
70. "Convertible Pyrolysis Tubes", Dohrmann Envirotech, Mountain View, Cal.
71. H. Landsberg and E. E. Escher, *Ind. Eng. Chem.* **46** (1954), 1422.
72. T-300-P Titration Cell, Instruction Manual, Dohrmann Envirotech, Mountain View, Cal.
73. D. F. Adams, G. A. Jensen, J. P. Steadman, R. K. Koppe, and T. J. Robertson, *Anal. Chem.* **38** (1966), 1094.
74. A. Cedergren and G. Johansson, *Talanta,* **18** (1971), 917.
75. F. C. A. Killer, "Sub-ppm sulphur analysis by microcoulometry". *In* D. R. Hodges (Ed.), "Recent Analytical Developments in the Petroleum Industry", Applied Science Publ., Barking, 1974.
76. H. V. Drushel, "Determination of sulphur below 1 ppm by microcoulometry", presented at Southwestern Regional ACS Meeting, Baton Rouge, La., December 1972.
77. "SBI Single Boat Inlet: Installation and Operation", Dohrmann Envirotech, Mountain View, Cal.
78. Application Note AI-27: "Solid/Liquid Pyrolysis Accessories for Microcoulometric Titrator", Dohrmann Envirotech, Mountain View, Cal.
79. "GLI Gas/LPG Sample Inlets: Installation and Operation," Dohrmann Envirotech, Mountain View, Cal.
80. Method ASTM D 3120-72 T: "Trace quantities of sulphur in light liquid

petroleum hydrocarbons by oxidative microcoulometry"; 1974 Annual Book of ASTM Standards, Part 25, Amer. Society for Testing and Materials, Philadelphia, 1974.

81. Method ASTM D 3246-73 T. "Sulphur in petroleum gas by oxidative microcoulometry"; *ibid.*, Part 25.
82. "Application Note MC-301: Trace Sulphur Analysis—Oxidative", Dohrmann Envirotech, Mountain View, Cal.
83. F. C. A. Killer, *J. Inst. Petroleum*, **59** (1973), 142.
84. F. P. Scaringelli, S. A. Frey, and B. E. Saltzman, *Amer. Ind. Hyg. Assoc. J.* **28** (1967), 260.
85. H. V. Drushel, "Microcoulometric determination of S,Cl, and N in petroleum fractions and GLC effluents", presented at Eastern Anal. Symposium, New York, November 1970.
86. E. Bremanis, J. R. Deering, C. F. Meade, and D. A. Keyworth, *Mater. Res. Stand.* **7** (1967), 459.
87. W. R. Bandi, E. G. Buyok, and W. A. Straub, *Anal. Chem.* **38** (1966), 1485.
88. J. S. Ball, "Determination of types of sulphur compounds in petroleum distillates", Bureau of Mines, U.S. Dept. of Int., R. I. 3591, 1941.
89. J. M. Carter, *Analyst*, **97** (1972), 929.
90. K. Hoshino, *Bunseki Kagaku*, **22**, (1973), 866; Chem. Abstr. 1974, 54598.
91. J. R. Glass and E. J. Moore, *Anal. Chem.* **32** (1960), 1265.
92. D. Fraisse and S. Raveau, *Talanta*, **21** (1974), 629.
93. Application Note MC-401, "Trace Sulphur Analysis—Reductive," Dohrmann Envirotech, Mountain View, Cal.
94. Application Note MC-402, "Platinum Pyrolysis Catalyst," *ibid.*
95. L. L. Farley and R. A. Winkler, *Anal. Chem.* **40** (1968), 962.
96. Application Note MC-201, "Trace Chlorine Analysis", Dohrmann Envirotech, Mountain View, Cal.
97. Application Note MC-101, "Trace Nitrogen Analysis", *ibid.*
98. L. A. Fabbro, L. A. Filachek, and R. L. Iannacone; R. T. Moore, R. J. Joyce, Y. Takahashi, and M. E. Riddle, *Anal. Chem.* **43** (1971), 1671.
99. H. E. Aavik, I. A. Revel'skii, A. V. Kabun, and A. Herem," Study and development of a method of microcoulometric detection of nitrogen-containing compounds", Tr. Vses. Issled. Soveshch. Ostatkov Pestits. Profil. Zagryazneniya Imi Prod. Pitan., Kormov Vnesh. Sredy, d. 1971, p. 28.
100. J. R. Spies and T. H. Harris, *Ind. Eng. Chem. Anal. Ed.* **9** (1937), 304.
101. D. R. Rhodes, J. R. Hopkins, and J. C. Guffy, *Preprints Div. Petr. Chem.* ACS, **13** (1968), 43; *Anal. Chem.* **43** (1971), 556.
102. I. J. Oita, *Anal. Chem.* **40** (1968), 1753.
103. I. J. Oita, *ibid.* **43** (1971), 624.
104. M. Kajikawa, M. Kawaguchi, T. Amari, and K. Watanabe, *Sekiyu Gakkai Shi*, **15**(9) (1972), 170; *Chem. Abstr.* 1973, 86759g.
105. E. Bishop, Coulometric Analysis. *In* C. L. Wilson and D. W. Wilson (Eds), "Comprehensive Analytical Chemistry," Vol. IID, Elsevier, Amsterdam, Oxford, New York, 1975.
106. J. S. Nader and J. L. Dolphin, *J. Air Pollut. Control Ass.* **8** (1959), 336.
107. H. C. MacKee and W. L. Rollwitz, *ibid.*, **8** (1959), 338.
108. P. B. Tarman, B. H. Andreen, and D. V. Kniebes, "Comparison of instrumental methods for odorants and other sulphur compounds in natural gas", Presented at Amer. Gas Assoc. Operating Section Conference, 1961.

109. J. Smejkal and J. Bartonek, *Pap. Celul.*, **30** (1) (1975), 27; *Chem. Abstr.* 1975, 144352g.
110. J. B. Pate, J. P. Lodge, Jr. and M. P. Neary, *Anal. Chim. Acta*, **28** (1963), 341.
111. D. F. Adams, W. L. Bamesberger, and T. J. Robertson, *J. Air Pollut. Control Ass.* **18** (1968), 145.
112. W. L. Bamesberger and D. F. Adams, "Automatic microcoulometric analysis of sulphur-containing gases in the atmosphere—interference and selective filter break-through time studies", presented at ACS 158th National Meeting, New York, September 1969.
113. W. L. Bamesberger and D. F. Adams, *TAPPI*, **52** (7) (1969), 1302.
114. A. E. O'Keeffe and G. C. Ortman, *Anal. Chem.* **38** (1966), 760.
115. A. Cedergren, A. Wikby, and K. Bergner, *ibid.* **47** (1975), 111.
116. P. Niklas, "Selective cartridges for removing interfering substances from a sulphur dioxide-containing sample gas stream", Ger. (East) Pat. 97,493, May 1973.
117. F. P. Scaringelli and K. A. Rehme, *Anal. Chem.* **41** (1969), 707.
118. E. Jacobsen and G. Tandberg, *Anal. Chim. Acta*, **64** (1973), 280.
119. W. Krijgsman, J. F. Mansveld, and B. Griepink, *Z. anal. Chem.* **249** (1970), 368.
120. A. Schait and J. Cuzner, *Amer. Soc. Brew. Chem. Proc.* 1970, 29.
121. N. Linnet, *Internat. Laboratory*, 1973 (Nov/Dec.), 25.
122. D. E. Ott and F. A. Gunther, *Bull. Environ. Contam. Toxicol.* **12** (1974), 161.
123. R. E. Rostenbach and R. G. King, *J. Air. Pollut. Control Ass.* **12** (1962), 459.
124. F. P. Scaringelli, J. P. Bell, and E. Rosenberg, "Determination of parts-per-billion concentrations of ammonia, nitric oxide, and nitrogen dioxide in air", Presented at ACS 158th National Meeting, New York, September 1969.
125. S. I. Krichmar, V. E. Stepanenko, and T. M. Galan, *Zh. Anal. Khim.* **26** (1971), 1840.
126. C. A. Boström, A. Cedergren, G. Johansson, and I. Pettersson, *Talanta*, **21** (1974), 1123.
127. A. Cedergren and G. Johansson, *Science Tools*, **16**(2) (1969), 19.
128. A. Cedergren, *Talanta* **21**, (1974), 367.
129. E. Barendrecht and J. G. F. Doornekamp, *Z. anal. Chem.* **186** (1962), 176.
130. R. A. Gibbs and R. J. Palma, *Anal. Letters*, **7**(3) (1974), 167.
131. R. S. Braman, D. D. DeFord, T. N. Johnston, and L. J. Kuhns, *Anal. Chem.* **32** (1960), 1258.
132. S. Miyake, *J. Japan Petroleum Inst.* **14** (1971), 975.
133. F. C. A. Killer, "Microcoulometric determination of traces of chlorine and sulphur in organic materials", Presented at 6th Internat. Symposium on microtechniques, Graz, Austria, September 1970.
134. J. Solomon and F. Uthe, *Anal. Chim. Acta*, **73** (1974), 149.
135. D. R. McFee and R. R. Bechtold, *Amer. Ind. Hyg. Ass. J.* **32** (1971), 766.
136. J. Sevcik, *Chromatographia* **4**(3) (1971), 102.
137. Application Note GT-101: "Trace Nitrogen Detector for the Detection of Nitrogen in Gas-Chromatographic Effluents", Dohrmann Envirotech, Mountain View, Cal.
138. Application Note GT-201, "Trace Chlorine Detector for the Detection of Chlorine in Gas-Chromatographic Effluents", *ibid.*
139. Application Note GT-301, "Trace Sulphur Detector (Oxidative) for the Detection of Sulphur in Gas-Chromatographic Effluents", *ibid.*

140. Application Note GT-401, "Trace Sulphur Detector (Reductive) for the Detection of Sulphur in Gas-Chromatographic Effluents", *ibid.*
141. B. Karlhuber, K. Ramsteiner, W. D. Hoermann and W. Simon, *J. Chromatog.* **84** (1973), 387.
142. V. T. Brand and D. A. Keyworth, *Anal. Chem.* **37** (1965), 1424.
143. M. Cranmer and J. Carroll, "Comparison of conductivity and microcoulometric methods for the detection of pesticides", Presented at ACS 154th National Meeting, Chicago, Ill., September, 1967.
144. W. E. Westlake, *Advan. Chem. Series*, 1971, 104; Pesticide Identification at the Residue Level, p. 73.
145. P. A. Mills, J. H. Onley, and R. A. Gaither, *J. Ass. Offic. Agr. Chem.* **46** (1963), 186.
146. H. L. Pease and J. J. Kirkland, *J. Agr. Food Chem.* **16** (1968), 554.
147. Y. Iwata, W. E. Westlake, and F. A. Gunther, "Evaluation of rapid procedure for determining organochlorine insecticides in raw and canned beef and cheese fat by microcoulometry-GC", Presented at 84th Annual Meeting, Association of Offic. Anal. Chemists, Washington DC, October 1970.
148. N. F. Wood, *Analyst*, **94** (1969), 399.
149. F. D. Griffith and R. V. Blanke, *J. Ass. Offic. Anal. Chem.* **57** (1974), 595.
150. H. E. Munro, *Pestic. Sci.* **3** (1972), 371.
151. H. L. Pease, *J. Agr. Food Chem.* **14** (1966), 94.
152. J. F. McCarthy, *Research Development*, 1970, 18.
153. D. F. Adams and R. K. Koppe, *J. Air Pollut. Control Ass.* **17** (1967), 161.
154. O. D. Smith and K. H. Tauss, *S. Pulp Paper Mfr.* **33** (2) (1970), 32, 34.
155. R. Robertus and M. J. Schaer, *Environ. Sci. Technol.* **7** (9) (1973), 849.
156. D. L. Brink, A. A. Pohlman and J. F. Thomas, *TAPPI*, **54** (1971), 714.
157. F. E. Saalfeld, F. W. Williams and R. A. Saunders, *Amer. Lab.* **3** (7) (1971), 8.
158. F. W. Williams, *Anal. Chem.* **44** (1972), 1317.
159. F. W. Williams and M. E. Umstead, *Anal. Chem.* **40** (1968), 2232.
160. F. C. A. Killer, *Proc. Soc. Anal. Chem.* **9** (1972), 71.
161. V. Zitko, *J. Chromatog.* **81** (1973), 152.
162. S. I. Krichmar and V. E. Stepanenko, *Zh. Analit. Khim.* **24** (1969), 1874.
163. V. E. Stepanenko and S. I. Krichmar, *ibid.* **26** (1971), 147.
164. D. J. David, "Gas Chromatographic Detectors", Wiley-Interscience, New York, 1974.

APPENDICES

I. DETERMINATION OF SULPHUR (OXIDATIVE METHOD)

I.1. Preparation of T-300-P titration cell

Check that both generator electrodes and the sensor electrode are perfectly clean. If in doubt, remove the cell cap carrying the sensor electrode and the generator anode from the cell (see Fig. 17), wash the electrodes with deionized water and acetone and dry. Heat the electrodes carefully to a bright orange colour in a Bunsen burner flame and allow them to cool slowly before immersion in the electrolyte. To clean the spiral generator cathode in the side arm, pour a small amount of warm aqua regia (hydrochloric acid: nitric acid 3:1) into the cell cavity (with the cell cap removed), allow it to drain into the side arm with the electrode and keep it there for 30 seconds producing a vigorous gas evolution on the metal. Wash the cell cavity and the sidearm thoroughly with deionized water to remove last traces of the acids.

To prepare the reference electrode, put approx. 2 g of resublimed iodine crystals in a small agate mortar, cover with electrolyte and grind coarsely, to a size of 20–40 mesh. Fill the cell to a height of ca. 5 cm with electrolyte and transfer the iodine in small portions to the electrolyte-filled reference sidearm, making sure that no air bubbles are trapped between the iodine granules. Insert the platinum reference electrode carefully into the iodine granules, with the ground joint lightly greased with silicone grease to ensure a tight seal. No air bubbles should be trapped in the electrolyte above the iodine.

Flush the cell with several volumes of electrolyte, making sure that no air bubbles are trapped in the cell, in particular in the generator and reference electrode sidearms. Place the stirring bar in the cell cavity, replace the cell cap and adjust the electrolyte level in the cavity to 5–10 mm above the electrodes. The cell is now ready for use.

I.2. Procedure

Choose a standard solution, the concentration of which is near the expected sulphur concentration of the sample and pull an adequate amount into the syringe (e.g. use a 10-μl syringe for 1–8 μl of solution). Withdraw the solution into the syringe barrel so that the lower meniscus falls on the 1-μl mark and record the liquid volume in the syringe. Repeat the same steps after injection; the difference between the readings is the volume of sample injected. Alternatively, the weight of sample injected may be established by weighing the syringe before and after injection. Insert the needle through the septum. A small peak may appear, particularly when working with low sulphur concentrations (at high range ohms). Wait until this "needle peak" is titrated and inject the solution at the proper injection rate.

The peak obtained will have one of the shapes shown in Fig. 22. If the peak has the correct shape B, the instrument is ready for the analysis of samples. Choose the range ohms so that a sizeable peak is obtained, if possible larger than half the recorder chart width. If the peak is overshooting (A), reduce the gain in small increments or reduce the bias voltage in 5-mV steps until the peak has the correct shape B. If the peak is tailing (C), increase the gain and/or bias voltage as described above to obtain the correct peak shape.

Inject the sample in the same way as the standard, using the same syringe and the same sample volume. Measure the peak area, preferably by means of an integrator. Repeat the injections until the repeatability of peak area counts is satisfactory. Inject standards at frequent intervals.

Check sulphur recovery periodically (e.g. every four hours) to be in the vicinity of 80%, by analysing a test solution with known sulphur content against a standard and by calculating the sulphur content of this solution from electrochemical data (see below). Check the blank of the microcoulometric system by periodical injection of sulphur-free isooctane or toluene. This blank which is usually less than 0·1 ppm sulphur should be subtracted from both sample and standard peak areas where it matters, i.e. when analysing samples containing sulphur at the 1-ppm level.

I.3. Procedure (Standby Technique)

Prepare the microcoulometric system for operation as described above with the function switch of the microcoulometer in the "operate" position (generating circuit switched on).

Adjust the plunger of a 25-μl microsyringe equipped with a 10-cm needle so that the tip is at the 2-μl mark. Draw 10 μl of sample into the syringe by raising the plunger tip to the mark which will give a total sample volume of 10 μl, including the sample in the needle. Retract the plunger further until the whole sample slug of 10 μl is in the syringe barrel, with a cushion of air above it and the needle empty. This technique avoids sample remaining in the syringe needle after the injection and thus reduces memory effects between two consecutive injections. Insert the needle through the septum and place the syringe on the cranking device. Wait until the needle peak is titrated.

Switch to the "standby" position (generating circuit switched off) and start a stopwatch at the same time. Using the cranking device, inject the sample at a uniform rate not exceeding 0·3 μl/s. The injection of 10 μl of sample takes about 30 seconds. Allow the cell to equilibrate for another 30–60 seconds and switch back to "operate". This will cause the appearance of a sharp peak (Fig. 24). Measure the peak area in the usual way.

Repeat the same procedure with an appropriate standard (e.g. 1·0 ppm thiophene or di-*n*-butyl sulphide in CFR grade isooctane) and with the pure iso-octane, always using the same duration of the combustion period (i.e. time on "standby") as for the sample. A total combustion time of 1–2 minutes is adequate. The time on "standby" must be kept constant to eliminate the effect of a slight baseline drift, caused by impurities in the gases, losses of iodine from the electrolyte, etc. Plot the peak area counts for standard and solvent versus their sulphur concentrations and read the concentration of the sample from the linear plot (Fig. 25).

II. DETERMINATION OF SULPHUR (REDUCTIVE METHOD)

II.1. Preparation of catalyst

Sieve 75 g of Norton SA-203 alundum support 10/20 mesh and digest for half an hour with boiling conc. hydrochloric acid. Wash the alundum with deionized water until free of chloride. Dry the washed alundum in air, place 75 g into a porcelain dish, add 23·5 g of $H_2PtCl_6 . 6H_2O$ and 30 ml of water and evaporate on a steam bath under constant stirring until the alundum is covered with a uniform red-brown coating. Condition the catalyst before use as described in Appendix II.2.

II.2. Preparation of pyrolysis tube

Pack the central section of the pyrolysis tube with the platinum catalyst, holding it in place between two platinum gauze baskets. To activate the catalyst before use, insert the tube into the furnace and bring the temperatures in the inlet section to $>750°C$ and in the centre and outlet sections to 1000°C. Connect the inert gas supply (argon or helium) to the pyrolysis tube and purge for at least 10 minutes at a flow rate of 200 cm^3/min to remove the air from the system. Replace the inert gas by hydrogen and condition the catalyst for at least one hour, using the same flow rate of hydrogen. The tube is now ready for use.[(94)]

II.3. Reconditioning of catalyst

Disconnect the titration cell from the tube outlet and attach a length of rubber tubing to the outlet in its place. Dip the other end of the tubing into a beaker filled with water. Replace the hydrogen supply to the pyrolysis tube by the inert gas (argon or helium) and purge the tube for at least 10 minutes at a flow rate of 200 cm^3/min. Watching the gas bubbling through the water in the beaker, replace the inert gas supply by oxygen and adjust the flow rate so that the bubbling continues. Initially, a very high flow rate will be required as the catalyst consumes oxygen. After a few minutes the flow rate can be reduced to 100 cm^3/min for the remainder of one hour. Replace the oxygen supply by the inert gas and condition the catalyst as described above.

II.4. Preparation of the T-400-S titration cell

Inspect the sensor electrode and the generator anode, both made of silver-plated platinum, for defects in the plating. If necessary, replate the electrodes as described in Appendix II.5. Clean the generator cathode made of a platinum spiral with aqua regia as described for the T-300-P cell (see Appendix I.1.). To prepare the reference electrode, fill the cell to a height of ca. 5 cm with the electrolyte, prepare a mixture of mercury and mercury(II) oxide and transfer it in small portions to the reference sidearm, making sure that no air bubbles are trapped. Insert the reference electrode with the ground joint lightly greased with silicone grease to ensure a tight seal. Flush the cell with several volumes of electrolyte, making sure that no air bubbles are trapped in the cell, particularly in the generator and reference electrode sidearms. Place the stirring bar in the cell cavity, replace the cell cap and adjust the electrolyte level in the cavity to 5–10 mm above the electrodes. The cell is now ready for use.

II.5. Silver plating of electrodes

Remove any old plating from the electrodes with hot nitric acid. Wash the electrodes thoroughly with deionized water and acetone and flame to bright orange colour, taking care not to damage the glass parts. Prepare the plating bath by dissolving 4·0 g of silver cyanide, 4·0 g of potassium cyanide and 6·0 g of potassium carbonate in deionized water and making the solution up to 100 ml. Filter the solution into a Pyrex container, where it will keep for approximately six months. Use analytical grade reagents.

Insert the clean electrodes into the plating bath and use a strip of analytical grade silver (dimensions $70 \times 3 \times 1$ mm) as the anode. Carry out the plating under the following conditions:

Reference electrode (wire)	2 mA, 45 min
Cell cap electrodes (simultaneously)	8 mA, 4 h
	or 2 mA, 16 h

After plating, soak the electrodes in acetic acid for 30 minutes and rinse with water.

II.6. Procedure

Select a standard solution the concentration of which is near the expected sulphur concentration of the sample and pull 1–8 µl into a 10-µl syringe. Withdraw the solution into the syringe barrel so that the lower meniscus falls on the 1-µl mark and record the liquid volume in the syringe. Repeat the same steps after injection; the difference between the readings is the volume of sample injected. Alternatively, the weight of sample injected may be established by weighing the syringe before and after the injection. Insert the needle through the septum. If a needle peak appears (particularly at attenuations set for concentrations below 5 ppm sulphur), wait until the peak is titrated and inject the solution at a rate not exceeding 0·2 μl/s. If the peak that appears has the correct shape (Fig. 22), the instrument is ready for the analysis of samples. Select the attenuation, amplifier gain, and bias voltage, inject the sample, measure the peak area, and repeat the injections, injecting standards at frequent intervals, as described in Appendix I.2.

Check the sulphur recovery periodically (e.g. every four hours) to be near 100%, by analysing a test solution with known sulphur content against a standard and by calculating the sulphur content of this solution from electrochemical data (see below). If the recovery drops below 95%, recondition the catalyst as described in Appendix II.3.

Check the blank of the microcoulometric system by periodical injection of sulphur free isooctane or toluene. This blank, which is usually less than 0·1 ppm sulphur, should be subtracted from both sample and standard peaks where it matters, i.e. when analysing samples containing sulphur near the 1-ppm level.

III. DETERMINATION OF CHLORINE

III.1. Preparation of T-300-S titration cell

Check the cell cap electrodes (sensor and generator anode) and the reference electrode, which are made of silver-plated platinum, for cleanliness and defects in the plating. If in doubt, replate the electrodes, using the silver plating procedure given in Appendix II.5. Do not attempt to clean these electrodes by flaming as this treatment will destroy the silver plating. If necessary, clean the generator cathode (platinum spiral) in the sidearm with aqua regia as described in Appendix I.1. and wash the cell thoroughly with deionized water to remove the last traces of the acids.

To prepare the reference electrode, fill the cell to a height of ca. 5 cm with electrolyte and pack the electrolyte-filled reference sidearm with silver acetate in powder form, making sure that no air bubbles are trapped in the layer. Insert the silver-plated reference electrode, with the ground joint lightly greased with silicone grease to ensure a tight seal. Flush the cell with several volumes of electrolyte, making sure that no air bubbles are trapped in the cell, in particularly in the generator and reference side-arms. Some silver acetate will pass through the perforated glass disc at the base of the reference sidearm which must be removed by tilting the cell and flushing with electrolyte. Place the stirring bar in the cell cavity, replace the cell cap and adjust the electrolyte level to 5–10 mm above the electrodes. The cell is now ready for use. Stability often improves after the electrodes have been allowed to equilibrate, e.g. after 24 hours. The equilibration period of freshly plated cell cap electrodes may be shortened by operating the cell at the usual conditions in the electrolyte (70% acetic acid) to which 250 μg of sodium chloride have been added.

III.2. Procedure

Select a standard solution, the concentration of which is near the expected chlorine concentration of the sample and pull an adequate amount into the syringe (e.g. use a 10-μl syringe for 1–8 μl of solution). Measure the volume or weight of the solution taken as described in the determination of sulphur by the oxidative method (Appendix I.2) and inject it into the combustion tube, taking care not to exceed an injection rate of 0·3 μl/s. If working at low chlorine levels, wait for the "needle peak" to titrate before injecting the solution in the syringe. If the peak has the correct shape (Fig. 28) the instrument is ready for the analysis of samples. Select the attenuation, amplifier gain, and bias voltage, inject the sample, measure the peak area, and repeat the injections, injecting standards at frequent intervals, as described in Appendix I.2.

Check the chlorine recovery periodically (e.g. every four hours) to be near 100%, by analysing a test solution with known chlorine content against a standard and by calculating chlorine content of this solution from electrochemical data (see below). Check the blank of the microcoulometric system by periodical injection of chlorine-free iso-octane or toluene. This blank which is usually less than 0·2 ppm chlorine should be subtracted from both sample and standard peak areas where it matters, i.e. when analysing samples with low chlorine contents.

IV. DETERMINATION OF NITROGEN

IV.1. Preparation of pyrolysis tube

Pack the central section of the pyrolysis tube with the nickel catalyst, holding it in place between two nickel gauze baskets. To activate the catalyst before use, insert the pyrolysis tube (without the scrubber tube) into the furnace and purge it of air by passing an inert gas (argon or helium) at 200 cm^3/min for at least 10 minutes. Replace the inert gas by hydrogen, adjust the flow rate again to 200 cm^3/min and raise the furnace temperatures to >400°C in the inlet section and to 800° in the centre and outlet sections. Activate the catalyst at these conditions for at least two hours. After this time, switch back to inert gas and purge for 10 mins at 200 cm^3/min to remove the hydrogen.

Connect the pyrolysis tube to the oxygen supply and attach a length of rubber tubing to the outlet of the tube. Dip the other end of the tubing into a beaker filled with water. Turn on the oxygen; watching the gas bubbling through the water in the beaker, adjust the flow rate so that the bubbling continues. Initially, the oxygen flow rate must be very high as the hot nickel metal consumes oxygen. After a few minutes the flow rate can be reduced to 100 cm^3/min and the purging continued for the remainder of one hour. Check the centre furnace temperature, as the oxidation is exothermic, and do not allow the temperature to exceed 850°C. Replace oxygen by the inert gas supply and purge at 200 cm^3/min for at least 10 minutes to remove oxygen from the pyrolysis tube. Replace the inert gas by the hydrogen supply and maintain a flow at 200 cm^3/min for at least two hours to complete the conditioning.

Adjust the furnace temperatures to those given in Table 17.

IV. 2 Reconditioning of catalyst

Disconnect the titration cell and remove the acid-gas scrubber. Establish the furnace temperatures specified above for the initial activation of the catalyst, pass hydrogen, and follow the entire procedure given above for the initial activation.

IV.3. Preparation of acid gas scrubber

Pack the tube to a length of 50–70 mm with the support material (Alundum chips 20/40 mesh, or pumice stone 16/30 mesh) and fix it in place with two pieces of nickel wire gauze. Add sufficient 50% NaOH to wet about 60% of the support. Insert the tube in the outlet section of the pyrolysis tube at 300°C with a hydrogen flow of 200 cm^3/min and dry for five minutes. Remove the scrubber tube from the pyrolysis tube, cool and repeat the addition of NaOH. Replace the scrubber tube and allow to dry.

IV.4 Preparation of T-400-H titration cell

Inspect the reference electrode (made of platinum plated with lead) and the sensor electrode (made of platinum covered with a platinum black deposit) for surface defects. If necessary, replate them as described in Appendices IV.5. and IV.6. The generator electrodes are made of bright platinum and can be cleaned as described in Appendix I.1.

To prepare the reference electrode, fill the cell to a height of ca. 5 cm with electrolyte and pack the electrolyte-filled reference sidearm with powdered lead sulphate, making sure that no air bubbles are trapped in the layer. Insert the reference electrode with the ground joint lightly greased with silicone grease to ensure a tight seal.

Flush the cell with several volumes of electrolyte, making sure that no air bubbles are trapped in the cell, in particular in the generator and reference sidearms. Any lead sulphate crystals that may have passed through the perforated glass disc at the base of the reference sidearm must also be removed by tilting the cell and flushing with electrolyte. Place the stirring bar in the cell cavity, replace the cell cap and adjust the electrolyte level in the cavity to ca. 10 mm above the electrodes. The cell is now ready for use.

IV.5. Lead plating of reference electrode

Remove any old plating from the electrode with a 50 + 50 (by volume) mixture of fluoboric acid (HBF_4) and hydrogen peroxide, and flame the electrode shortly at a dull orange heat, taking care not to heat the glass parts. Fill a 10-ml plastic beaker with the plating solution (lead fluoborate, 50% w/v), and immerse the reference electrode and a lead strip (70 × 3 × 1 mm) or rod (diameter 3–4 mm) into the electrolyte to serve as the anode. Take care that the electrodes do not come in contact with each other. Attach the electrode to the cathode and the lead strip to the anode of the plating unit and pass a current of 50 mA for two minutes.

Inspect the electrode; if it shows any flaws (e.g. exposed platinum), repeat the entire procedure. Plate until a smooth grey plating is obtained, wash the electrode with water and rinse with acetone.

IV.6. Deposition of platinum black on sensor electrode

Remove any old platinum black coating from the electrode by short flaming at dull orange heat, taking care not to heat the glass parts. Fill a small beaker with a 5% solution of chloroplatinic acid ($H_2PtCl_6 \cdot 6\,H_2O$) and immerse the cell cap electrodes in the plating solution. Attach the sensor electrode to the cathode of the plating unit and the generator anode to the anode, and pass a current of 10 mA for 25 seconds. Do not stir the bath during plating. Wash the electrodes with water and rinse with acetone. If the platinum black coating appears to be non-uniform, repeat the entire procedure.

IV.7. Procedure

Select a standard solution, the concentration of which is near the expected nitrogen concentration of the sample and pull an adequate volume into the syringe (e.g. use a 10-μl syringe for 3–8 μl of solution. Measure the volume or weight of a standard solution in the usual way and inject it into the pyrolysis tube, taking care not to exceed an injection rate of 1 μl/s. If working at low nitrogen levels, allow the "needle peak" to titrate before injecting the solution in the syringe.

If the nitrogen peak has the correct shape (see Fig. 29), the instrument is ready for the analysis of samples. Select the attenuation, gain, and bias voltage, inject the sample, measure the peak area, and repeat the injections, injecting standards at frequent intervals, as described in Appendix I.2.

Check the nitrogen recovery periodically (e.g. every four hours) by analysing a test solution with known nitrogen content against a standard and by calculating the nitrogen content of this solution from electrochemical data (see below). Recoveries of less than 90% are suspect and may be due to catalyst deactivation, excessive coke formation or scrubber tube inefficiency.

Check the blank of the microcoulometric system by periodical injection of nitrogen-free toluene or iso-octane. The blank which is usually less than 0·1 ppm of nitrogen should be subtracted from both sample and standard peak areas when analyzing samples with low nitrogen contents.

Check the performance of the acid gas scrubber tube periodically (e.g. daily) by injecting 5 μl of acetone under normal operating conditions. If CO_2 breaks through, indicated by a negative peak, and continues to do so after several such injections, add a little more NaOH, as described in Appendix IV. 3, or replace the scrubber.

IV.8. Procedure using the boat inlet

With argon flowing through the boat inlet and the pyrolysis tube, replace the septum inlet with the boat inlet. Switch from argon to hydrogen, with dry hydrogen at 200 cm^3/min passing through the boat inlet and humidified hydrogen at 200 cm^3/min flowing directly into the pyrolysis tube. Adjust the furnace temperatures as follows:

Inlet section	1000°C
Centre	750°C
Outlet	300°C

Attach the titration cell and push the empty boat into the inlet section several times, until the peak obtained is reduced to a negligible size (i.e. less than 1% of the sample peak).

Position the boat below the injection port (see Fig. 20) and inject a sample. Before withdrawing the syringe touch the tip of the needle against the inside of the boat to complete the transfer of sample. Advance the boat to the hydrogen inlet to the pyrolysis tube. After evaporation of the solvent is completed, advance the boat into the centre of the inlet section. When the titration is complete, withdraw the boat to the heat sink for 30 seconds, then return to the position under the injection port. If after the titration the peak does not return to the original baseline, discard the result and re-zero the recorder before withdrawing the boat. Repeat the injections. Run standards after the first sample, then after every three or four samples.

INDEX